Italian Soft-Skinned Vehicles of the Second World War
Motorcycles, Cars, Trucks, Artillery Tractors 1935–1945

Volume 1

Ralph Riccio

Mario Pieri

Daniele Guglielmi

Helion & Company

Helion & Company Limited
Unit 8 Amherst Business Centre
Budbrooke Road
Warwick
CV34 5WE
England
Tel. 01926 499 619
Email: info@helion.co.uk
Website: www.helion.co.uk
Twitter: @helionbooks
Visit our blog at blog.helion.co.uk

Published by Helion & Company 2023
Designed and typeset by Farr out Publications, Wokingham, Berkshire
Cover designed by Paul Hewitt, Battlefield Design (www.battlefield-design.co.uk)

Text © Ralph Riccio, Mario Pieri, Daniele Guglielmi 2023
Photographs © see Acknowledgements
Colour artwork by and © David Bocquelet 2023

Every reasonable effort has been made to trace copyright holders and to obtain their permission for the use of copyright material. The author and publisher apologise for any errors or omissions in this work, and would be grateful if notified of any corrections that should be incorporated in future reprints or editions of this book.

ISBN 978-1-804513-27-9

British Library Cataloguing-in-Publication Data.
A catalogue record for this book is available from the British Library.

All rights reserved. No part of this publication may be reproduced, stored in a retrieval system, or transmitted, in any form, or by any means, electronic, mechanical, photocopying, recording or otherwise, without the express written consent of Helion & Company Limited.

For details of other military history titles published by Helion & Company Limited contact the above address, or visit our website: http://www.helion.co.uk.

We always welcome receiving book proposals from prospective authors

Contents

List of colour plates	iv
Acknowledgements	v
Documentation sources	vi
Authors' Notes	vii
Abbreviations	viii
Glossary	ix
Foreword	xi
Prefixes and suffixes used in vehicle designations	xii
Introduction	13
1 Overview and Explanatory Notes	17
2 Motorcycles	30
3 Motor Cars	55
4 Light trucks	98

A bibliography will appear in Volume 2

List of colour plates

Fiat 508 CM (aka Fiat 1100 Mimetica) staff car, Italy, 1939. (Artwork by and © David Bocquelet)	I
Autocarretta 35 light truck, mountain chain of the Alps, Italy, 1935. (Artwork by and © David Bocquelet)	I
Fiat 618 light truck, East Africa, 1936. (Artwork by and © David Bocquelet)	II
SPA 38R light truck, Eastern Front, 1941. (Artwork by and © David Bocquelet)	II
SPA CL 39 light truck, Eastern Front, 1941. (Artwork by and © David Bocquelet)	III
SPA AS 37 (second series) light truck, North Africa, 1942. (Artwork by and © David Bocquelet)	III
Benito Mussolini and Marshal Pietro Badoglio on board a Bianchi VM6 C staff car in a newsreel frame from 28 June 1940.	IV
Photograph taken by a German soldier during the transfer trip and stay in North Africa. On the left, an Opel Olympia 38 that appears camouflaged in the *Tropen* colours employed by the Wehrmacht since March 1942. On the right, a Fiat 508 C Camioncino belonging to the *Regio Esercito* whose grey-green colour is only partially covered by the *kaki sahariano* (sand-yellow) applied by spray gun.	IV
A 508 L Camioncino, painted grey-green.	V
An advertising poster for the Breda company (1931).	V
The cover of a 1934 motorcycling magazine with Sertum company propaganda.	VI
The original brand of coachbuilder Viberti.	VI
The updated version of the Viberti brand.	VI

Acknowledgements

The authors and the publisher would like to give thanks to the following individuals for the precious collaboration granted in the creation of this book: Thomas Anderson, Massimo Bartolini, Luigi Carretta, Flavio Chistè, Stefano Di Giusto, Enrico Finazzer, Andrea Olivero, Claudio Pergher, Alberto Pirella, David Zambon, Ennio Zanetti. In particular, we are grateful to the Gruppo Modellistico Trentino di Studio e Ricerca Storica (www.gmt.tn.it – info@gmt.tn.it) and Claudio Pergher again for providing photographs, drawings and other material from its publications and to Flavio Chistè for his constant assistance.

Documentation sources

The documents that appear in this volume come from the repositories and archives cited in the text, or from the authors' private archives. The authors are willing to settle any form of copyright issues pertaining to images and documents, the original provenance of which it has been impossible to determine.

ACS	Archivo Centrale dello Stato, Italy
AUSSME	Archivio Ufficio Storico Stato Maggiore Esercito, Italy
BAMA	Bundesarchiv, Militärarchiv, Germany
CSM	Centro Studi della Motorizzazione, Italy
CTM	Centro Tecnico della Motorizzazione, Italy
ECPAD	Établissement de Communication et de la Production Audiovisuelle de la Défense, France
LIFE	*Life* magazine, USA
GMT	Gruppo Modellistico Trentino di Studio e Ricerca Storica, Italy
ISR	Istituti Storici della Resistenza, Italy
IWM	Imperial War Museum, UK
MMM	Museo della Motorizzazione Militare, Italy
MSGR	Museo Storico Italiano della Guerra di Rovereto, Italy
NARA	National Archives and Records Service, USA
USMM	Ufficio Storico della Marina Militare, Italy

Authors' Notes

The importance of land transport vehicles within an armed force is often underestimated by the average reader, attracted by more 'martial' subjects such as tanks and artillery. Nevertheless, it was thanks to motorcycles, cars, trucks and tractors that – since the early years of the twentieth century – men, weapons, ammunition, provisions, fuel, equipment and orders were transported, all elements without which AFVs, guns and infantry are unable to fight.

In this book we focus on the means of transport in force in the Italian Royal Army (and, in some cases, also in the Italian Royal Air Force and Navy) from the 1930s to the end of the Second World War. Little has been said about them in recent years, even in Italy, with some exceptions such as the Guzzi Alce motorcycle, the Fiat 508 CM car, the Fiat 626/666 and Lancia 3Ro trucks, and a few light and medium tractors.

It is common opinion that the Italian Army was beaten above all because of the poor quality of its combat vehicles. Actually, impartial and in-depth studies made since shortly after the end of the war, have revealed that the main problem was the shortage of vehicles, as well as an entirely insufficient logistics chain. The tank crews were able to compensate with bravery and experience for the fact that their tanks were, from a certain point on, inferior to those of their enemies, but the inadequate number of AFVs and other material was impossible to remedy. The same problem plagued the entire sector of military soft-skinned vehicles, a sign of Italy's limited industrial capacity (and rather of procurement of raw materials and components) compared, for example, to its ally Germany. There were too few factories, too few skilled workers and poor management skills within the armed forces.

However, if quantity was lacking, the same cannot be said for quality. Many models of efficient, robust and resistant vehicles were produced, especially in the sector of the so-called 'standardised' motor vehicles, such as those mentioned at the beginning and others that we will see. These vehicles allowed Italian troops to move and fight in the large and difficult territories of North Africa, the Balkans and the Soviet Union and which brought home what was left of the defeated soldiers.

Abbreviations

AOI	Africa Orientale Italiana (Italian colonies in East Africa)
AS	Africa Settentrionale (North Africa)
MdS	Milizia della Strada, Milizia Nazionale della Strada (National Road Militia)
MVSN	Milizia Volontaria per la Sicurezza Nazionale (Volunteer Militia for National Security)
PAI	Polizia dell'Africa Italiana (Italian African Police)
RA	Regia Aeronautica (Royal Italian Air Force)
RCTC	Regio Corpo Truppe Coloniali (Royal Corps of Colonial Troops)
RCTL	Regio Corpo Truppe Libiche (Royal Corps of Libyan Troops)
RE	Regio Esercito (Royal Italian Army)
RM	Regia Marina (Royal Italian Navy)

Glossary

Italian terms
Aerflex: very low pressure 'balloon' pneumatic tyre for cars and light vehicles
Artiglio: pneumatic tyre with large treads ('artiglio' literally means 'claw')
Autobus, autocorriera: bus
Autocarro: truck, lorry
Autocarro unificato medio: standardised medium truck
Autocarro unificato pesante: standardised heavy truck
Autogruppo: transportation battalion, consisting of several *autoreparti* (transportation companies)
Autoraggruppamento: echelon consisting of several *autogruppi* (transportation battalions); transportation brigade
Autoreparto: transportation company, consisting of several *autosezioni* (transport sections)
Autosezione: motor transport section
Autovettura: motor car
Autovetturetta: compact motor car
Biposto: two-seat motorcycle
Camion: truck, lorry
Camioncino: small truck
Camionetta: light truck
Carrello elastico: a single-axle two-wheel bogie trailer on which antiquated artillery pieces were placed in order to make them suitable for high-speed towing
Carro: initially the term indicated a truck or lorry; later it was used for the *carro armato* (tank) and therefore the truck became *autocarro* to avoid misunderstandings
Celerflex: a type of semi-pneumatic tyre
Cingolato: tracked
Cingoletta: light tracked vehicle
Coloniale, Col.: modified versions of a civilian or military vehicle in order to operate in desert and tropical environments, typical of Italian African colonies
Cord: standard pneumatic tyre for light and heavy vehicles
Furgoncino: small van
Furgone: van
Grigioverde: standard grey-green colour for Italian Army materials in the European theatre
Kaki sahariano (or *giallo sabbia*): standard sand-yellow colour for Italian Army equipment in the North African theatre from middle of 1941
Leggero: light
Medio: medium
Militare, Mil.: military
Milizia Marittima di Artiglieria (MILMART): naval artillery militia, normally coastal artillery, but also manned truck-mounted naval guns
Milizia Nazionale della Strada (MdS): branch of the *Milizia Volontaria per la Sicurezza Nazionale* with traffic police functions
Milizia Volontaria per la Sicurezza Nazionale (MVSN): Volunteer Militia for National Security, a Fascist militia organisation
Modello, Mod.: model
Monoposto: single-seat motorcycle

Motocarrozzetta: motorcycle/sidecar combination
Motocicletta, motociclo: motorcycle
Mototriciclo: three-wheeled motorcycle
Pesante: heavy
Polizia Coloniale: original name of the *Polizia dell'Africa Italiana*
Polizia dell'Africa Italiana (PAI): Italian African Police, the police corps of Italian North Africa and Italian East Africa colonies from 1936 to 1945, reporting directly to the Ministry of the Colonies, later renamed the Ministry of Italian Africa. Until 1939 the corps was named *Polizia Coloniale*
Regia Aeronautica (RA): Italian Royal Air Force
Regio Corpo Truppe Coloniali (RCTC): Royal Corps of Colonial Troops, Italian military unit which included troops stationed in the African colonies of Eritrea, Somalia, Tripolitania and Cirenaica (Tripolitania and Cirenaica were later merged into Libya)
Regio Corpo Truppe Libiche (RCTL): Royal Corps of Libyan Troops (1938–1943); formerly Regio Corpo Truppe Coloniali della Libia (1935–1938)
Regio Esercito (RE): Italian Royal Army
Regia Marina (RM): Italian Royal Navy
Rimorchio: trailer
Semicingolato: half-track
*Superflex (*aka *Superflex Cord):* low pressure pneumatic tyre for light and heavy vehicles
Sigillo Verde: special tread for heavy vehicles tyres suitable for soft soils
Sigillo Verde Libia (aka *Tipo Libia*): special tread for tyres designed for sand
Stella Bianca: special tread for light vehicles, similar to the *sigillo verde*
Trattore: tractor
Trattrice: tractor (usually heavy)
Ultraflex: very low pressure 'balloon' pneumatic tyre for heavy vehicles

Foreword

I am grateful to write this section for the book, *Italian Soft-Skinned Vehicles of the Second World War: Motorcycles, Cars, Trucks, Artillery Tractors 1940–1945* by Ralph Riccio, Daniele Guglielmi and Mario Pieri. I feel as such, not only because it is an honour to be asked by these well-respected authors, but also because this book is needed. There is very little English language reference material covering these subjects which is available to military vehicle enthusiasts and model hobbyists.

Since first becoming interested in military vehicles as a boy, my primary interest for the past twenty years has become armoured cars and wheeled fighting vehicles. Although related, I must admit I was not terribly predisposed towards soft-skinned vehicles. However, over the years as I became more knowledgeable about armoured cars, my interest could not but be drawn towards the types of vehicles covered in this book. The obvious automotive background and technical similarities shared by both armoured cars and soft-skinned vehicles almost made it a necessity that I focus more on the latter, to better understand the former.

During my experiences, needless to say, I have found that there is a vast amount of printed and online references available covering armoured fighting vehicles. Although, I have not been terribly disappointed in my search for information covering soft-skinned vehicles, it is harder to come by than with other some other vehicles. In addition, as one will recognise quickly while learning or researching military history in general, and equipment and vehicles specifically, some countries will often be the subject of more works than others. Unfortunately, Italy seems to be one of the nations covered less than others, especially by English language sources. This situation is even more skewed when it comes to Italian military vehicles, specifically soft-skins. As a result, I have found it difficult to locate high-quality English language reference material covering this subject.

So needless to say, as a military enthusiast and model hobbyist, I am very pleased that this book has been written by these noted authors and published for our use. It is a welcome addition to my reference collection. I also can assure you, it won't be collecting dust on my shelf, as I will be referencing it often.

Patrick Keenan
Editor – WarWheels.net
September 2023

Prefixes and suffixes used in vehicle designations

AS	Autocarro sahariano (desert truck)
B	Benzina (petrol, gasoline)
BM	Benzina, Militare (military vehicle with petrol engine)
C	Coloniale (tropicalised vehicle or equipment)
C	Corto (short wheelbase)
CL	Carro leggero (light truck)
CM	Carro militare (military truck)
CV	Carro veloce (fast tank, actually a tankette)
G	Gassogeno (gas generator engine)
GM	Gassogeno, Militare (military vehicle with gas generator engine)
L	Lungo (long wheelbase)
M	Militare (military vehicle)
N	Nafta (diesel oil)
NM	Nafta, Militare (military vehicle with diesel engine)
P	Pneumatici (pneumatic tyres)
PC	Pesante Campale (heavy field)
R	Ribassato (lowered chassis for use for buses and other special vehicles)
S or SP	Semipneumatici (semi-pneumatic tyres)
TL	Trattore leggero (light tractor)
TM	Trattore medio (medium tractor)
TP	Trattore pesante (heavy tractor)

Introduction

Part (perhaps a very large part) of the reason that the *Regio Esercito* did not perform as well as it might have otherwise have done during the Second World War was that it lacked sufficient mobility, compromising its ability to perform competitively on the battlefield against opponents that were more highly motorised and mechanised. Limited mobility was a direct result of the fact that the Italian automotive sector was never able to provide the number of trucks required to support the armed forces, especially the *Regio Esercito*. Despite the fact that Italian infantry divisions were much smaller than those of the other major combatant nations (the standard infantry division was a 'binary' division consisting of only two, rather than three, infantry regiments), most Italian infantry divisions were chronically short of trucks, especially enough trucks to move the troops, and barely had enough trucks and tractors to equip the artillery regiments within the division. As early as the opening moves of the desert war between Italy and Britain, prior to September 1940, Marshal Rodolfo Graziani, who was acutely aware of the deficit of motor transport in North Africa, pleaded with Mussolini for 600 additional trucks so that he could completely motorise his attack force. Mussolini denied the request and ordered that the planned attack proceed without the trucks. The situation did not improve as the war progressed: in July 1941, there were only about 5,200 trucks available to an Italian force in North Africa consisting of 110,000 men.

The reasons for this woeful shortage of vehicles were many and varied; foremost among these reasons was that Italy's industrial base was simply too small to be able to adequately provide all of the vehicles required to equip an army consonant with the times. One of the endemic problems with the Italian industrial base that produced equipment for the military forces was that the too few companies were spread too thin, producing hardware for all three services; as examples, Fiat built aircraft, tanks and motor vehicles while Ansaldo built ships and artillery among other items. Other factors that affected production

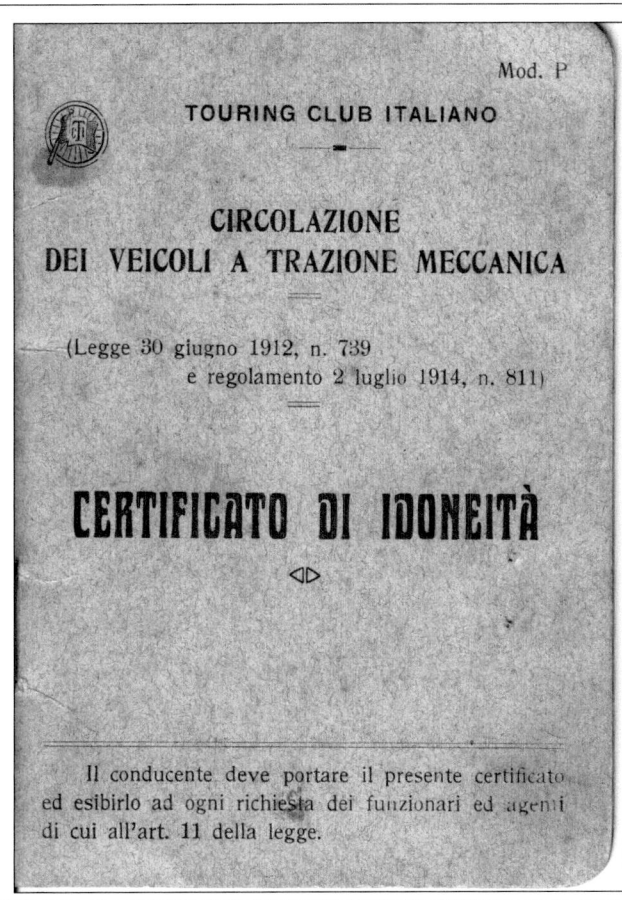

An Italian driving licence assigned on 26 January 1916 to a Corporal of the *87° Reggimento Fanteria* of the *Regio Esercito*.

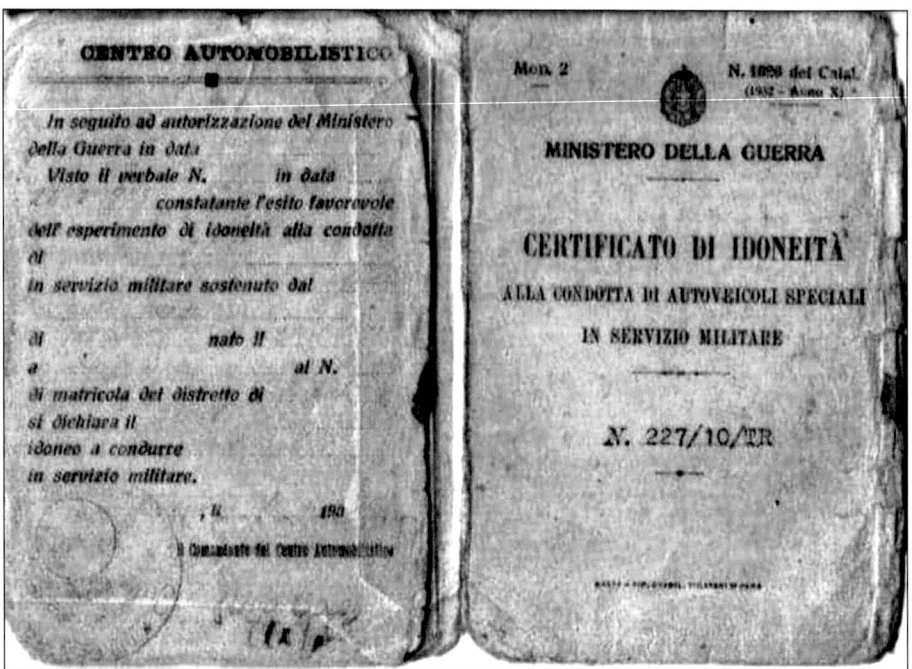

A blank driving licence issued by the War Office in 1932.

The heterogeneity of the fleet of logistic vehicles used by the Italian Army is exemplified by these trucks of an Autosezione marching in Libya in the winter of 1940–1941. From left: an old Fiat 618, a Fiat 634 N2 and a SPA Dovunque 35. (ACS)

Fiat 500 Topolino under construction in the Fiat Mirafiori plant, which had been opened in 1939. (Fiat)

were a scarcity of primary materials to support the entire war effort and the lack of an orchestrated effort to coordinate and maximise existing production capabilities.

Among the automotive companies suitable for the Italian military, the high-volume producers (in a relative sense) were Fiat, and its subsidiaries SPA and Ceirano. Compared with this 'giant consortium', Lancia, Breda, Isotta Fraschini, Officine Meccaniche (OM) and Bianchi were somewhat lower-volume car and truck manufacturers. In addition to producing cars and trucks (both civilian and military), the Fiat group also produced tanks, engines and aircraft. Likewise, Alfa Romeo built aero engines, Isotta Fraschini built aero engines and marine engines for motor torpedo boats as well as anti-aircraft guns, and Breda produced railway equipment, and light weapons and anti-aircraft guns. Almost inevitably, and not surprisingly, when needs were prioritised, soft-skinned motor vehicles came in towards the bottom of the pile. And as if that wasn't enough, these companies were in competition and rarely worked together. The General Staff of the Italian Army calculated in 1938 that the aforementioned companies could monthly produce 650 heavy trucks, 600 light trucks, 300 vehicles of various types (mainly tractors), 100 *autocarrette* and 50 heavy tractors, for a total of 1,700 motor vehicles, to be reduced to 1,500 considering the needs of the other Italian armed forces. Actually, these figures were achieved only for short periods.

The result was that, although tanks and guns may have taken priority over trucks, without trucks (and tractors) the artillery and its ammunition could not be moved, the tanks could not be transported to where they were needed, and the infantry would have to sit where it was and could not be moved in time

A column of motor vehicles of the Italian *Regio Esercito* moving towards the front lines of the Eastern Front through a town in the summer of 1941. (ACS)

to influence the course of a battle. Given the scarce resources at hand, the Italian military and industry faced an impossible dilemma when it came to trucks or combat equipment. In oversimplified terms, if Italy produced a greater of number of trucks at the expense of combat equipment, there would have been little or no equipment such as artillery to transport, whereas if the priority had been given to major combat items (as was the case), there would not be enough trucks, or tractors, to adequately support the combat arms.

Perhaps one bright spot in the Italian vehicle production picture was the number of motorcycles the various Italian manufacturers were able to produce to meet military requirements. Motorcycles required relatively few raw materials and did not consume much fuel per vehicle, but the downside was that a motorcycle could carry only one or two soldiers, which meant that it took approximately ten to twenty motorcycles to carry as many men as a single truck, somewhat negating the lower cost and fuel savings per motorcycle.

Although an imprecise yardstick, it is illustrative to consider a few figures regarding Italian truck production compared to production in a similar timeframe in Germany and the United States. During the Second World War, Italy had an army of roughly 4 million men (ground forces only, excluding navy and air forces), while Germany had around 13 million and the USA around 11 million, so both Germany and the USA had approximately three times as many soldiers as Italy. As the following illustrates, even accounting for the relative sizes of the armies involved, there was nevertheless a huge disparity between Italian truck production and that of Germany and the USA.

Before the outbreak of the Second World War, according to official sources the Italian Army had about 22,000 motor vehicles in service (6,487 light trucks, 5,887 heavy trucks, 2,332 *autocarrette*, 2,921 special vehicles, 4,328 tractors). By June 1940 the total had risen to 53,000 vehicles, excluding tanks, of which 17,000 were requisitioned in previous years. In June 1943 the armed forces (Army, Navy and Air Force) had 96,000 vehicles and 40,000 motorcycles, most of them manufactured in Italy.

Nicola Pignato (see bibliography) estimated that between 1940 and 1945, Italy produced about 163,200 motor vehicles (cars, trucks and tractors). By comparison, between 1935 and 1944, Germany produced

Type of truck	Load		Speed in kilometers per hour	
	Men	Material (kg)	Single vehicles	Columns (average)
Light:				
Infantry Truck	10 to 12	1,000	40	25
OM4 OMF	12 to 15	1,200	60	30
Fiat 618 CM	12 to 15	1,250	65	30
Spa 25 C 10	16 to 20	1,800	50	25
Fiat 612 P	20 to 25	2,500	43	25
Spa 38 R	20 to 25	2,500	52	25
Ceirano 47 CM	20 to 25	3,000	45	25
Bianchi Mediolanum 36	20 to 25	3,000	55	25
OM, CRD	20 to 25	3,000	51	25
Isotta Fraschini D 70 NM	20 to 25	3,000	56	25
Heavy:				
Ceirano 50 CM	20 to 25	5,000	25	18
Lancia RO.NM	24 to 30	5,000	32	20
RO BM	24 to 30	5,000	39	22
Fiat 633 NM	24 to 30	5,000	30	20
Fiat 633 GM	24 to 30	3,500	28	18
Isotta Fraschini D 80 NM	24 to 30	5,000	34	20
OM 3 BOD	24 to 30	5,000	51	23
Giant:				
Fiat 634 N	28 to 32	7,000	40	20
3 RO Lancia	28 to 32	6,500	43	22

Table of motor transport loads, from *TME 30-420 Handbook on the Italian Military Forces* (3 August 1943).

approximately 130,000 Opel Blitz trucks alone, while the figures for US truck production are simply staggering in comparison. US automotive manufacturers churned out more than 200,000 Studebaker US 6 trucks (the US 6 was actually manufactured by REO as well as by Studebaker, both relatively small companies by American standards) most of which were provided to the Soviet Union, as well as 572,500 GMC CCKW-series 2 half-ton trucks and 382,350 Dodge WC-series three-quarter-ton trucks, not to mention 647,925 Willys MB and Ford GPW jeeps – almost two million vehicles, excluding special vehicles and heavy 6x6 trucks built by manufacturers such as Diamond T, White, International Harvester, Autocar, Mack, Brockway, Corbitt, FWD and Ward La France which numbered well into the hundreds of thousands.

In the years leading up to the Second World War, the British government prepared the industry for the mass production of required motor vehicles. Britain entered the war with 80,000 military vehicles of all types, although most were left behind in the evacuation at Dunkirk in 1940. The Canadian auto industry not only replaced these losses, it did much more, producing more than 800,000 military transport vehicles.

Unlike its German ally which plundered all of the territories it occupied for anything related to the war effort, including requisitioning of motor vehicles as well as of the factories that produced them (Renault, Citroën and others in France; Tatra and Skoda in Czechoslovakia), the countries that Italy was able to occupy during the early stages of the war (Albania, Greece, the Balkans, East Africa and North Africa) yielded no such windfalls. Requisitioning civilian vehicles in the Italian colonies or Italy itself did little to fill the needs of the military; in 1940, a plan by the army to requisition 20,500 civilian trucks fell short by 7,900. It is useful to note that the availability of vehicles in Italy prior to the war was quite limited compared to the vehicle numbers registered in nearby countries; in 1939, Italy had only about 300,000 cars, or about one for every 130 people, whereas neighbouring France had almost 2 million cars, or one for every 23 people.

Historians have discussed a lot about the reasons why a direct and fruitful collaboration between the German and Italian factories never materialised, despite the alliance between the two nations and the proclamations of the two leaders and their main collaborators. Basically, as Lucio Ceva and Andrea Curami have described (see bibliography), the main shortcoming was on the Italian side. Professional jealousies, fear of losing commercial influence, political and industrial manoeuvres meant that, for example, no German engine of adequate power to be able to move tanks and heavy vehicles arrived in Italy, let alone finished products. At the same time, only a few vehicles requisitioned by the Germans in the occupied territories were forwarded to the *Regio Esercito*.

It is also necessary to underline the Italian war effort in the wars of the thirties in East Africa and (even if not officially) in Spain. These commitments drained the nation's military funds and reduced the circulating fleet of transport vehicles, as the army resorted to purchased or requisitioned civilian vehicles for reasons of time and resources.

The dire shortage of trucks in the Second World War was further compounded by their performance characteristics, which by comparison fell below those of some of the US trucks mentioned above. The GMC CCKW was a 6x6 truck with a 91.5 HP engine, a range of 300 miles (482km), and a road speed of 45mph (72km/h), while Italian trucks such as the Fiat 626 and Bianchi Miles had engines of about 65 HP, had a range of 400km (249 miles) and a road speed of about 63km/h (40mph). In fact, as mentioned above, the low power and, at times, reliability of some Italian engines, especially diesel engines, was a very significant 'Achilles' heel'. This in no way detracts from the overall quality of Italian trucks in general, but the lower performance characteristics vis-à-vis trucks of some other nations ultimately translated to even less ability to move troops and equipment as expeditiously as needed.

1

Overview and Explanatory Notes

Historical context

It can be said that, during the 1930s, the interest of the Italian Army was concentrated on two aspects: the defence of the country's land territorial borders, made up entirely of the mountain chain of the Alps, and the conquest and/ or administration of colonial territories in North and East Africa.

In the first case, it was evident that in the event of a war the main effort would have been performed by infantry, mountain troops and artillery, so the military authorities decided to adopt a light and very mobile tank – essentially a tankette – such as the Carro Veloce CV 33 (later the CV 35 and CV 38 versions), to be used as a fire support system. The army also equipped itself with light trucks (called *autocarrette*) with a maximum speed of only a few km/h but theoretically capable of traversing the mule tracks in the mountains, and wheeled artillery tractors suitable for towing field and siege guns. In the colonies, however, armoured cars, including those

An *autocarretta* (light truck) while facing a steep slope during the summer manoeuvres of 1932.

A column of military motor cars, headed by a Lancia Aprilia Coloniale, together with a sidecar and other Italian vehicles in a North African town. (D. Zambon)

Fragment of an advertising poster from 1939. The caption explains: 'FIAT 666 N. The new heavy-duty truck. Standardised type. It will surpass the success of its famous predecessor: the 634 N'.

that had survived from the Great War, were more useful, together with motor vehicles for the transport of men and materials.

With regard to the latter category, procurement was based on what was offered by domestic manufacturers. Only towards the end of the decade did the so-called *coloniale* (colonial) versions of many cars and trucks appear; these were modified versions based on standard models, intended both for the civilian and military market in order to operate in the desert and tropical environments typical of Italian African colonies.

At the time, Italian industry was still able to offer models of good overall quality, sturdy and quite reliable; at least as regards their use on the national territory, rich in hills and mountains. In spite of this, the army had mostly old and very diverse vehicles in active service, such as the Fiat 18 BLR and Fiat 15 Ter trucks. Italy's victories in the Italo-Ethiopian War (1935–1937) and in the Spanish Civil War (1937–1939) hid the existing problems relating to the numbers of vehicles available. In 1939 there were fewer than 50,000 examples serving in the army: old, new, civilian and specifically military. In fact, when Italy's entry into the war alongside Germany in June 1940 required a greater number of motor vehicles – not just AFVs – Italian industry revealed its limits. The shortage of qualified personnel (also because many men had been called up to the armed forces), the difficulty in finding materials and components, the scarcity of public funds because much had been spent in previous wars, the rivalry between companies, and finally political intrigues were among the main causes behind a production that was much lower than needed.

Campaigns of the Italian armed forces in North Africa and the Soviet Union, still alongside their German ally, in addition to operations in the Balkans, often put the entire logistics chain in crisis. Thanks to the self-denial and sacrifice of many crews and soldiers, the army still managed to fight.

After the Armistice of 8 September 1943, the Germans took over the entire Italian war production, controlling the factories. It is obvious that any type of vehicle was precious at the time; however Mussolini's former ally benefited from the availability of many Italian cars, trucks and tractors as it continued its war against Allied troops.

Autocarri Militari Unificati (Standardised Military Trucks)

During the war in Ethiopia, the Italian armed forces used a considerable variety of motor vehicles, including those of foreign manufacture. The vehicle fleet existing in Italy at that time was varied and, in the event of general mobilisation, it would have been possible to requisition many thousands of cars and trucks of different types and brands. This would perhaps have solved

A service station in 1935. (Fisogni Museum)

The first Fiat diesel engine for trucks, the four-cylinder model 350 with direct injection of 5,570cc mounted on the Fiat 632 N. (Negri Foundation)

the urgent needs, but on the other hand complicated the logistic situation even more.

Therefore, in July 1937 a decree was issued to concentrate production as much as possible, reducing the number of civilian vehicles belonging to the categories of light, medium and heavy trucks. The Italian manufacturers were told what common features these vehicles should have, in terms of engines, transmissions, drive wheels, cabs, performance, fuel capacity, weights; they would have 18 months to adjust production, otherwise the vehicles would not be admitted into circulation. In November of the same year, with the publication of the official rules, the concept of the *automezzo unificato* (standardised) motor vehicle was thus born.

The new classification of vehicles was based mainly on the engine type, the total weight and the maximum load. Newly built 2 axle trucks had to fall into one of the following three types:

1. heavy trucks with diesel engine, maximum laden weight of 12,000kg and maximum load of at least 6,000kg;
2. medium trucks with a maximum laden weight of 6,500kg and maximum load of at least 3,000kg;
3. light trucks with a maximum laden weight not exceeding 4,000kg and maximum load of at least 1,500kg.

This categorisation received some changes over time.

Even the main mechanical parts – from bolts to leaf spring leaves for the suspensions – were to have been interchangeable. The standardisation also covered towing pintles, braking and electrical systems, tyres and headlights. Also recommended was the adoption of diesel engines instead of gasoline engines, with low revs and high torque values. The placement of door windows in canvas and mica or celluloid instead of glass had to be provided for. It is possible that the decision, although logical in itself, was inspired by the analogous programme of Italy's German ally who, a short time before, had imposed rules for the production of *Einheits* type (standardised pattern) vehicles, with simplified parts.

Due to the participation of the Italian Expeditionary Forces in the Spanish Civil War, there was neither time nor opportunity for manufacturers to put the recommendations into practice, and this further revealed the complications that existed in managing such a heterogeneous fleet. As in other cases, in fact Italian industry adapted very slowly to the directives of the military authorities, fearing that much of its production could be requisitioned rather than sold. This situation persisted until the outbreak of the Second World War, so much so that when Italy entered the war on 10 June 1940, standardisation was still a long way off.

In any case, some progress had been made: the models in production were fewer than a few years before and had similar characteristics. However, it should be noted that, despite the use of old motor vehicles – even veterans of the Great War – or requisitioned civilian vehicles, as well as those purchased abroad or war booty once hostilities began, the needs of the Italian armed forces were never satisfied, leaving an endemic shortage that had serious consequences on the mobility and efficiency of the troops on all fronts. Most of the units were not independent in terms of logistics vehicles and had to use special motorised detachments (named *autoraggruppamenti, autogruppi, autoreparti* and *autosezioni*, at different levels) at the disposal of the armies and army corps; but they never had a sufficient amount of equipment to transport men and materials as required for military operations. Many trucks and their drivers worked tirelessly, until a mechanical failure, an accident or enemy action stopped them.

From the chassis of the standardised trucks, many derivatives such as buses, vans, tank trucks, ambulances, radio vehicles and mobile workshops could be obtained. Normal traction was on a single drive axle, the rear one, even though the design of four-wheel drive vehicles was encouraged – a feature much requested by the military units.

These standardised vehicles were mainly cars and trucks, but there were also trailers and motorcycles; they were widespread during the Second World War and in the years following. The standardised trucks included in the medium weight category the Fiat 626, perhaps the most famous, together with Bianchi Miles, Alfa Romeo 430, OM Taurus, Isotta Fraschini D65, and Lancia Esaro (this latter only entered production in mid–1943). The heavy category included the Fiat 666, Lancia 3Ro, Isotta Fraschini D80, Alfa Romeo 800, and OM Ursus.

Engines and fuels

Since the end of the nineteenth century, internal combustion engines were widespread in Italy for civil, industrial, agricultural and military uses. It was, after all, two Italians Niccolò (Eugenio) Barsanti, a teacher of mathematics and physics born in Pietrasanta in 1821, and Felice Matteucci, an engineer born in Lucca in 1808, who were the first in the world to design and patent an internal combustion engine. It was 1854 and two years later they built a prototype in Florence that used a mixture of air and illuminating gas (illuminating gas was made from bituminous coal and was widely used for municipal lighting).

Refuelling of Fiat Dovunque and other trucks in Libya at the end of 1940. (ACS)

Refuelling Fiat 634 N2 trucks belonging to the *271ª Autosezione*. (ACS)

In the early twentieth century, gasoline was the most widely used fuel in Italy for transport vehicles, or more precisely the only one. For example, a manual published in 1922 explains that benzene was not widespread, like ethyl and methyl alcohol, even when mixed with the former. On the other hand, diesel engines were bulky and noisy at the time, so their use was reserved for ships, industrial plants and, later, locomotives.

Only at the beginning of the 1930s engines fuelled by *nafta* (the term used for diesel fuel) began to be mounted on heavy civilian and military trucks, once they could be made smaller and more manageable. In summary, the first diesel engines for vehicles had limited power and their starting was complex. On the other hand, their efficiency was higher than gasoline-powered engines; this allowed lower consumption and, consequently, greater range.

After years of study, Fiat installed the first diesel engine for trucks on the 632 N model (civilian) in 1931; the first Fiat military truck running on diesel was the 633 NM introduced in 1935. Other companies, on the other hand, preferred to build engines of foreign manufacturers under licence, such as OM (under licence from Switzerland's Saurer) and Isotta Fraschini (under licence from Germany's MAN).

Military authorities began to prefer diesel engines, even for soft-skinned vehicles, because they are less liable to flammability, an especially important factor in combat vehicles. In this sense, the important parameter is the 'flash point', which for gasoline is about -20°C (-4°F): this implies that at room temperature it spontaneously emits vapours which are visible to the naked eye and are liable to catch fire with a minimum ignition, i.e. a heat source that provides the thermal energy. Diesel fuel, on the other hand, has a flammability temperature of about 55°C (131°F), so under normal environmental conditions it is much less prone to ignite than gasoline: only if it is heated to above 55°C does the emission of flammable vapours occur. Moreover, the mechanical efficiency was higher with a diesel engine, with a lower fuel consumption.

In Italy, until the outbreak of the First World War, there were few motor vehicles in use; refuelling was done by buying gasoline in various places, such as pharmacies and drugstores. The first fuel pumps were installed where the number of

A column of Lancia 3Ro; the leader is a bowser (tanker truck) and has the cab roof painted with the Italian tricolour flag for aerial recognition.

vehicles was greater, such as in garages, bus terminals and in some luxury hotels. After the end of that conflict, civilian motor vehicles in use in Italy numbered just over 30,000, a third of which were trucks. In WW1 the army operated just over 2,000 cars, all of civilian production, and about 500 specialised vehicles (ambulances, tank truck, buses), as well as a few thousand motorcycles used mainly for communications; on the other hand, the number of trucks exceeded 40,000 and of artillery tractors over 900 (without taking into account the losses).

Italian jerry cans equip a SPA AS 42 Sahariana; their resemblance to the German model is evident.

Gradually, in the twenties, on the basis of first American, then French and German experiences, the first gasoline pumps – called metering pumps – were built and approved, to be placed along some roads and motorways. The following decade was characterised by an economic crisis and, consequently, by a drop in consumption, but with the spread of diesel engines for trucks, even the distribution of fuels saw an evolution. Soon, street filling stations dispensed both petrol and diesel fuel.

Beginning from the first wars waged by Italy in the African territories and then in Spain in the second half of the thirties, the problem of how to transport fuels was a concern of the military authorities. However, for years 200 litre drums were the only container used by the army to supply transport, reconnaissance and combat vehicles. Only towards the middle of 1941, during the North African campaign against the British forces, was the jerry can built on the German model (*Einheitskanister*) adopted. It differed from that of Italy's German ally in having

An Italian jerry can built on the model of an early German type. The white painted 'X' shaped notch indicates that it contains potable water. (ACS)

Italian equipment captured by the Germans after the Armistice of 8 September 1943. Behind are a Guzzi Alce motorcycle, a number of jerry cans and some Fiat 626 trucks. (ECPAD)

the letters 'R.E.' (more rarely, with the letters 'RA' and 'RM') and the wording *20 litri* (20 litres) imprinted on the left side. Jerry cans marked with a white stripe or cross indicated that the contents were drinking water; however, often, as veterans tell us, the water was not drinkable because out of necessity containers that had previously been used to transport petrol or diesel were used.

Tanker trucks (bowsers) made their appearance during the war, often requisitioned civilian vehicles, and 200 litre fuel drums also remained in use.

Wheels and Tyres

The military specifications for acceptance required, at least until the mid-thirties, the presence of puncture-resistant tyres, since punctures were frequent because of the conditions of the roads, which were often scattered with nails lost from the metal shoes of draft animals, and – in the combat zone – from shrapnel, splinters and small arms bullets. At the beginning, leaving out the all-metal wheels and other solutions for tractors, solid rubber ring tyres were chosen. The downside was their rigidity, which made the crew's ride uncomfortable and tiring, also causing vibrations and stresses to the mechanical parts, eventually damaging them. Additionally, solid tyres had a tendency to overheat.

A partial solution to the problem was found with semi-pneumatic tyres, which included one or more cavities, called ventilation chambers, containing air at ambient pressure. In this way they were softer and less prone to overheating. The most modern anti-perforation semi-pneumatic model used by the Italian armed forces was the 'Celerflex', produced by Pirelli from 1937 on the basis of the experiences gained in previous years. Another was the 'Cellastico', made in Italy by Società Italiana Industria Gomma under licence from a Dutch manufacturer supplying various European armies.

Over time, the semi-pneumatic tyres gradually gave way to pneumatic tyres. Military authorities sometimes required a vehicle to fit both types. In fact, the negative aspects of semi-pneumatic tyres mentioned above were highlighted in the military campaigns of the 1930s in North and East Africa and in Spain; for this reason, they were abandoned. At the outbreak of the Second World War all the vehicles of the *Regio Esercito* were fitted with pneumatic tyres, although various photographs from the period show older vehicles still equipped with semi-pneumatic tyres, especially in the European theatre of war.

Semi-pneumatic tyres were also placed on the wheels of the

A 1926 advertising poster for Pirelli semi-pneumatic tyres, 'ideal for heavy vehicles'.

The chassis of a TM 40 medium tractor with Celerflex semi-pneumatic tyres.

carriages of artillery pieces. Cycles and motorcycles, on the other hand, were equipped only with pneumatic tyres.

The rubber tyres used by the Italian military vehicles of the time were, with rare exceptions, all manufactured by the Milanese company, Pirelli, which also had a solid civilian market outside Italy. Its catalogues, of course, varied over time, introducing new models with different construction features, dictated by technological advancement or by contingent needs. Pirelli tyres were designed for both civil and military use, since the armed forces also used vehicles not specifically designed for military purposes. Only after the introduction of standardised vehicles and trailers was there some rationalisation from a military point of view.

Two different Pirelli semi-pneumatic tyres.

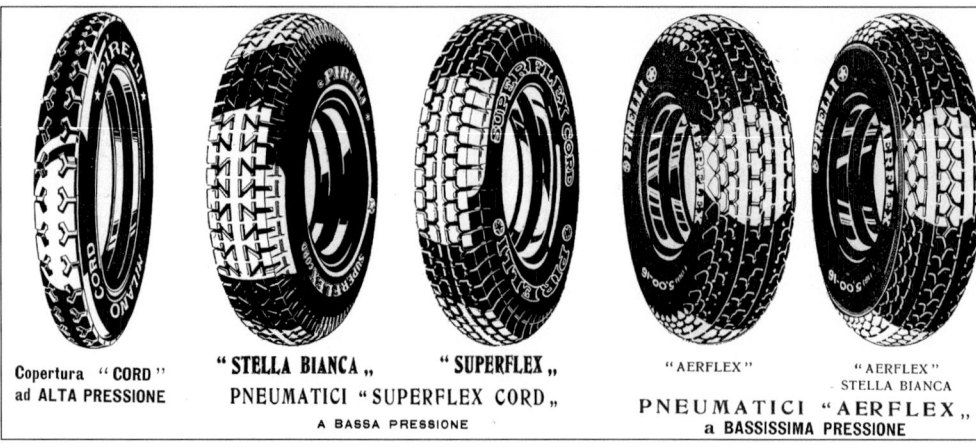

Five different tyres for motor cars and light vehicles illustrated on the Pirelli price list dated 1933. From left: the Cord, also available with artiglio tread (not shown here); the Superflex Cord with increased cross-section and low pressure, in the variants Stella Bianca and normal Superflex treads; the very low pressure Aerflex, also in this case including 'Normal' and 'Stella Bianca' variants.

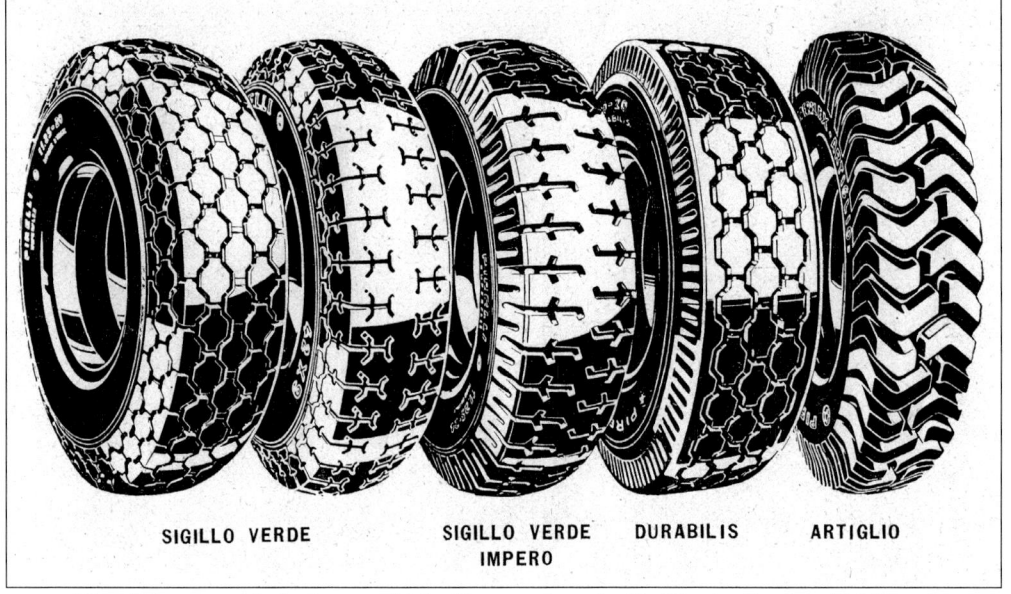

Five different treads for heavy vehicles illustrated on the Pirelli price list dated 1940 and available on the Cord, Superflex and Ultraflex tyre models. From left: the *sigillo verde* in two variants, the first one with the characteristic polygonal mesh engravings, while the second variant is the Raiflex; the Sigillo Verde Impero, featuring the 'f' carvings; the Durabilis for long-distance vehicles; finally the most recent Artiglio tread.

Strictly speaking, we should examine the tyres of all categories of motor vehicles in service in the 1930s and 1940s in the Italian armed forces. However, the list would be long, so we will limit ourselves to mentioning the most popular models and their treads, reiterating that there was no distinction between civil and military models.

In the older models of tyres, the carcass, that is the flexible and inextensible casing that contains the air chamber, was built – in simple terms – by immersing several layers (plies) of knitted cotton fabric in the melted rubber; in this way, the flexibility of the rubber was associated with the resistance to stresses given by the fabric. During the first and second decades of the twentieth century, the cord fabric (short for the French *corduroy*) was introduced; it replaced the common transverse cotton fabric as it is more durable and less subject to heating from friction. The Pirelli model Cord of 1921 represented the type of standard tyre for this manufacturer for many years.

In 1924 Pirelli introduced a new model, the Superflex Cord (or, more simply, Superflex), characterised by a larger section and low pressure inflation, followed later (early thirties) by the Aerflex and the Ultraflex at very low pressure inflation; the Aerflex was intended for passenger cars and light trucks, the Ultraflex for heavy vehicles. A lower pressure corresponded to a lower propensity to punctures, greater comfort and better grip on less compact ground, thanks also to the wider section: tyres of this type were also universally called *balloon*. High-pressure and low pressure Cord and Superflex Cord tyres were also available for two-wheeled and three-wheeled motorcycles.

Towards the end of the decade, these basic models were produced in the Raiflex variant, which had the carcass with a rayon fabric, a synthetic textile fibre derived from cellulose; in this way the Italian industry tried to reduce its dependence on imports, in this case of cotton. In addition, rayon guaranteed greater uniformity and resistance thanks to the structure of its fibre.

An aesthetic and functional differentiation among tyres was given by the tread, that is the part that makes contact with the ground.

One of the most common treads for the aforementioned Pirelli 'flex' models was the *Sigillo Verde* (Green Seal) patented in 1930; the manufacturer offered it in different sizes suitable for both wheels for cars or light vehicles and for heavy transport vehicles (in this case they were called 'giant tyres'). The sigillo verde tread was lightly sculpted, with a pattern that varied depending on the model of tyre that it was on. Apparently the most widespread pattern – at least for military vehicles – had almost hexagonal

A photograph from the *Salone dell'Automobile* (Motor Show), held in Milan in 1939. The Stella Bianca and Sigillo Verde tyres are displayed on the Pirelli stand. (Pirelli Foundation)

grooves, while other patterns featured diverse shapes.

Although documentation is quite sparse, it seems that at least initially the Superflex sigillo verde made of rayon (Raiflex) had a tread with grooves roughly in the shape of an 'H'.

In conjunction with the entry into force of the regulations concerning standardised military vehicles (issued in 1937, but implemented only over the next two to three years), a further variant of the sigillo verde appeared, called sigillo verde impero. In this tread the pattern had grooves in the shape of a lowercase 'f'. The sigillo verde impero pattern for Superflex and for Ultraflex tyres was highly employed in medium and heavy civilian trucks.

The other famous tread on tyres of Italian military vehicles, in particular tractors, armoured cars and vans, was the *artiglio* (claw), roughly comparable to the US NDT (non-directional tread) and the British 'cross-country' military tyres. From its launch in 1932, several variants were produced, outwardly similar to each other; there were, for example, semi-pneumatic and 'flex' artiglio tyres, with high and low inflating pressure. The artiglio tread was very sculpted and, according to the manufacturer, was suitable 'for heavy vehicles, intended for routes on sand, snow and marshy ground'; it was also designed to 'avoid the use of chains'. Actually, field experience showed that it tended to 'dig' into soft ground such as sand and mud until the vehicle bogged down; chains were still necessary in the snow.

For this reason, at the beginning of 1940 with the imminent opening of a war front in Africa, Pirelli put on the market a variant of the Superflex sigillo verde tread, for large-diameter wheels only (9.75 x 24 or 11.25 x 24 tyres – see below), such as to reduce the ground pressure, while decreasing the angle of incidence between the wheel and the sandy ground. The Superflex Libia, as it was called, had a great success even though production was never enough to equip all of the vehicles operating in North Africa.

The *stella bianca* (white star) tread for cars and light vehicles was the 'father' of the Sigillo Verde, having been conceived in 1927. The pattern of stella bianca for the Superflex family displayed characteristic 'Z' grooves, while the same for ultraflex was different and was also called *lusso* (luxury).

Particularly from 1942, due to the exhaustion of natural rubber stocks, tyres began to be manufactured in synthetic

The Superflex Sigillo Verde with the standard polygonal mesh pattern of tread.

The best known pattern for the Stella Bianca tread.

Two images of another Superflex Sigillo Verde, but with the tread pattern of the Raiflex variant. On the side you can read Pirelli Superflex and 11.25 – 24, with the star seal of Pirelli in the middle.

The Raiflex tread pattern for heavy vehicles featured 'H' shaped carvings, as in the case of the Cord 42 x 9 visible in section on the left. (Archivio Storico Fondazione Pirelli)

A Superflex Sigillo Verde Impero. Also in this tyre you can see 'Pirelli Superflex', the star, and '11.25 – 24'.

rubber, with a notable decline in quality and durability. However, after 8 September 1943, the Germans removed the machinery in use in Italy to continue production in Germany.

The two reference measures reported by the technical manuals are the nominal width of the tyre, and the diameter of the rim on which the tyre is mounted. However, the nomenclature was not unique.

In some cases, the two values were expressed in inches and separated by a dash, a slash or 'x' sign: 5.50 – 18 (but also 5.50 x 18) meant a 5½ inch width and a rim diameter of 18 inches.

In particular for the semi-pneumatic tyres, the two measures were instead expressed in millimetres, as for 175 x 720, that is a 175mm width and a rim diameter of 720mm.

There were, mainly in low and very low pressure tyres, mixed indications in millimetres and inches: 210 x 20 (but also 210 – 20) meant a 20-inch diameter rim and a 210mm width. Sometimes the two measures were written in reverse order.

Finally, it should be noted that usually the sidewall of the tyre had approximately the same size as the width.

Motorcycles had classic wire wheels. Special wire wheels were also fitted to some tractors. The wheels of the other motor vehicles were made of steel, essentially of two categories. The first category was made up of stamped discs, made of steel or other alloys, which could have circular or oval shaped lightening holes; often the front and rear wheels were interchangeable. Usually, the disc was bolted to the hub to facilitate disassembly.

The second category, which came into service in the second half of the 1930s and specifically for heavy vehicles, included sturdy wheels in cast steel characterised by six or eight large hollow spokes, forming a single body with the hub. The Gianetti company manufactured them in Italy having obtained the licence from the American Dayton Steel Foundry in 1932. In some texts this model of wheel is indicated with the denomination 'artillery

As you can read on its side, this tyre is a Pirelli Cord 32 x 6 artiglio early type.

The artiglio tread pattern varied over time and depending on the tyre model. At least three types are recognised.

A Superflex with artiglio tread. Note the different pattern compared with the 'Cord Artiglio'.

A Pirelli Superflex Libia 9.75 x 24, as can you read on its side.

type' inherited from the first wooden spoke wheels.

The rim, if not integral with the wheel disc, could be one-piece or divided in several parts. In particular, the Trilex model rim (born from a 1936 patent of the German company Georg Fischer) consisted of three segments fixed to the internal disc of the wheel with clamps; in twin wheels (double truck tyres) an intermediate external ring served as a spacer.

Semi-pneumatic tyres could be fitted to both categories of wheels but were preferably mounted on disc wheels. Cars and light trucks were equipped with disc wheels, with or without lightening holes. Trucks and tractors had their tyres usually mounted on spoked wheels.

The weakness of the tyres and the poor condition of the roads encountered in all operational theatres caused many punctures, for which there were not always enough spare tyres carried on the vehicles themselves, forcing repairs to be made continually, most of which did not last for long.

Electrical Systems

Even at the end of the thirties not all vehicles of the Italian armed forces were equipped with a battery, so the presence of a crank

A 20 inch spoked wheel Dayton-type manufactured by Gianetti. It consisted of a hexagonal-shaped cast steel sheet with six hollow spokes having holes at the ends for fixing to the rim; another six holes were present along the sides of the hexagonal disc (in this picture screws, bolts and other accessories are inserted). In the centre was the hub cap. (Museo delle Industrie e del Lavoro del Saronnese)

The effects of extended travel in the Libyan desert on the wheels of a Fiat 634. The outer tyre is a Pirelli Superflex with Sigillo Verde tread, the innermost with a much worn 'Sigillo Verde Impero' tread. (ACS)

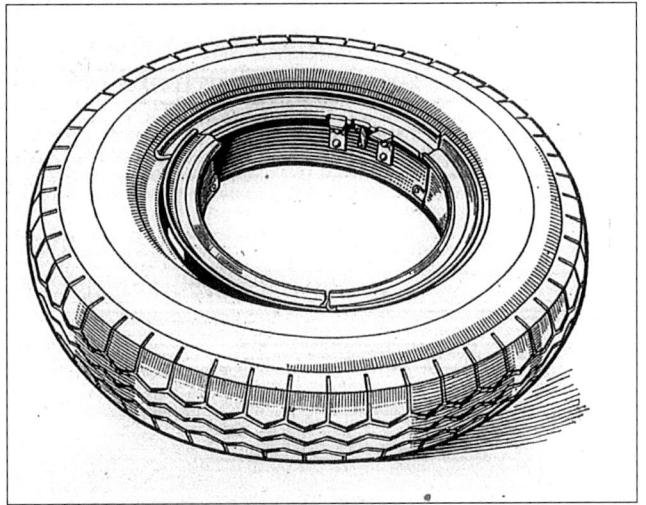

The structure of a 20 x 6 wheel with Trilex rims and tyres mounted on Fiat 626 trucks.

for starting the engine was an essential requirement. Since, as a result, the electric starter was also missing, an inertia starter could be present on heavy trucks and tractors; the case of the famous Lancia 3Ro is typical.

For the same reason, i.e. the lack of battery, other requirements for military trucks called for the presence of non-electric emergency headlights (petroleum, acetylene or carbide headlamps).

Right-Hand Drive

It might seem strange or puzzling to both British and American readers that Italian trucks were fitted with right-hand drive, in view of the fact that Italy practised the European custom of driving on the right-hand side of the road, although the legislation in this regard only entered force at the end of 1923 and became operational gradually.

The reason for this apparent anachronism lies in the fact that well into the 1920s and 1930s, outside of cities themselves, there were very few paved roads in Italy; this was also the reason that for many years the Italian Army requested that, in order to avoid punctures, trucks be fitted with semi-pneumatic rather than pneumatic tyres. Many roads were typically very narrow and shoulders were either nonexistent or were not very stable, neither were there any guard rails. The risk of ending up in a ditch or hitting a tree close to the side of the road was high; many trees close to the road had part of their trunks painted white precisely to act as a warning to drivers. Because of these reasons, having the steering wheel on the right-hand side enabled the driver to better judge how close the truck was to the edge of the road or to trees or other obstacles (large rocks, bridge parapets or, in some cases, even buildings) and thus avoid accidents. The steering wheel on the right-hand side of the vehicle made overtaking difficult, but heavy trucks were travelling at very low speeds and therefore overtaking other moving vehicles was infrequent. During the post-war years, Italian trucks transitioned to left-hand drive to conform to Continental driving practices.

Types of Bodywork

For the motor cars of the time, the two most common bodywork configurations, as for military purposes, were called in Italian 'torpedo' and 'berlina'.

The 'torpedo' body style, a tourer in England, was characterised by a completely open passenger compartment, with half-doors without windows, located immediately behind the engine compartment. The engine hood and the sides of the passenger compartment were equalised so as to present an almost continuous surface, we could say 'aerodynamic'. At the rear, the bodywork

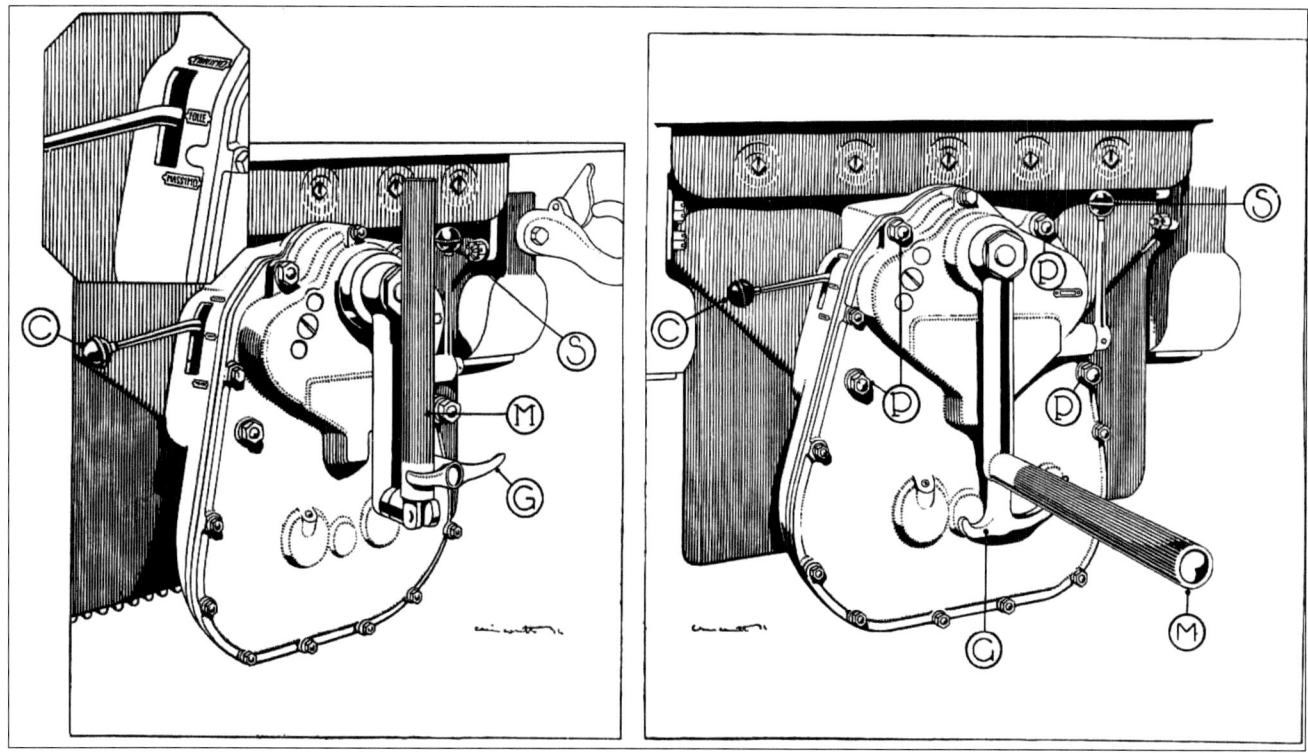

The inertia starter of the Lancia 3Ro truck.

ended with the passenger compartment itself, the tail being absent or at most limited to a small luggage compartment. There was usually a soft top, customarily in canvas, rigid or folding.

The 'berlina' body style, sedan or saloon in English, was and still is the classic closed body, with a fixed rigid roof, two-box or three-box design and, usually, with four or more doors.

To these two must be added the 'spider', the roadster in English, a sports body style for a two-seater, convertible car.

As far as heavy vehicles are concerned, the truck (otherwise called a lorry) was equipped with a rear bed, while the van has a completely closed body. It is appropriate to mention the types of cab. In the *cabina avanzata*, cab over engine (COE) in English, cab and engine are located above the front axle, therefore the front of the vehicle and the windshield are aligned, and the vehicle is 'flat nosed'. In the *cabina arretrata*, in the American cab – or conventional cab in England – the cab is located behind the engine and the nose of the vehicle is prominent. Finally, the *cabina semi-avanzata* is a compromise between the previous ones, in which the windshield is aligned with the front axle and the nose is small.

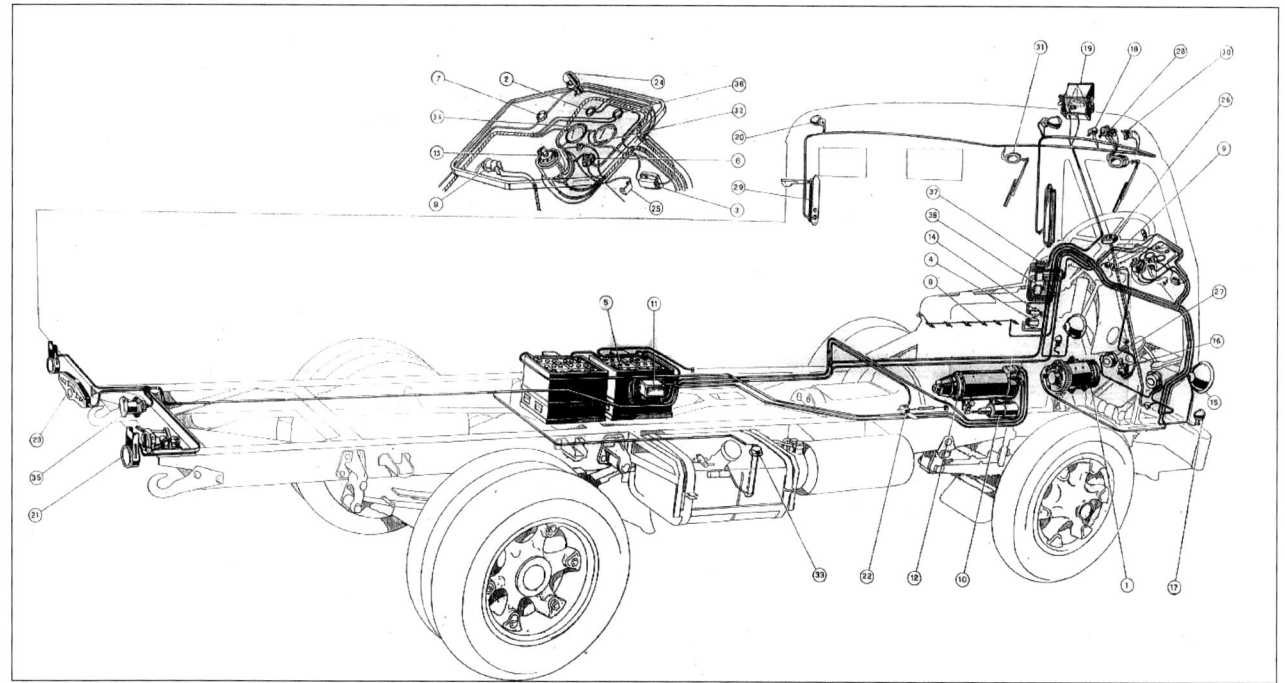

Electrical system of the Fiat 626 NLM truck.

2
Motorcycles

Motocicletta Volugrafo Aermoto

Despite the worn negative, the photo shows the overall structure of the small Aermoto.

German soldiers driving the Aermoto in Italy.

An example examined by British soldiers in Rome, in via Casilina, in June 1944.

Developmental and Service History

The firm Officine Meccaniche Volugrafo, based in Turin, produced fuel gauges for airplanes in the 1930s, which were also sold to the Luftwaffe. Volugrafo can in fact be translated as a volume indicator. It had already submitted some experimental models of vehicles to the Italian Army, without success.

The Volugrafo Aermoto Aviolanciabile was a parachutable light motorcycle based on a 120cc engine and designed in 1936 by engineer and company owner Claudio Belmondo. It was the forerunner of the post-war Vespa and Lambretta motor scooters, however, it was not itself successful following the war. Starting in 1938, the Italian armed forces were setting up the first airborne units and studying how to make paratroopers mobile once they were on the ground.

The Volugrafo Aermoto was accepted as a small parachutable motorcycle in conjunction with the planning for Operation C3, the invasion of Malta, which did not in fact occur. The final model was ready in 1942 and a first series of machines were produced in 1943. An initial issue of 2,000 examples was scheduled for the 183rd Paratroopers Division 'Ciclone', with Aermotos also being issued to the parachute school at Tarquinia and the 'San Marco' Regiment of the Italian Navy. However, the vehicle was never used en masse and many examples remained in the army's warehouses. Following the September 1943 Armistice, production of the Volugrafo Aermoto continued for a few months under German supervision and some examples were assigned to German paratroop units along the Adriatic coast and in the Rome area.

Technical Description

The Volugrafo Aermoto had many innovative features, especially for such a small machine. The frame was a rigid double cradle made of tubular steel, and exhaust gases were vented through the tubular frame itself, causing a lot of smoke and noise. The handlebars could be folded so that the machine could fit into a container for dropping by air; the specially designed container measured 1,300mm (51 inches) long by 800mm (31 inches) diameter and was cushioned internally with duck fabric. Once the container had landed, the Volugrafo Aermoto could be removed from the container, have its handlebars repositioned and locked, and be ready for action in two minutes. The gas tank was under the seat. The two-stroke engine was derived from the Sertum Batua 120 motorcycle running on a mix of petrol and oil. During the design phase, it was realised that the vehicle had to have wheels of a very small diameter but be capable of withstanding the weight of a man and his equipment; since suitable wheels and tyres did not exist, the choice fell on trolley wheels which were placed double on each axle; however, the motorcycle always tended to travel on the right or left wheels. After the war the Volugrafo Aermoto was modified and equipped with a single wheel. Brakes were drum brakes. The machine could tow a small trailer that had an 80kg (176lbs) carrying capacity.

Variants

A three-wheeled experimental variant, designated as a *motocarrello per aviotruppe* (motorised cart for airborne troops) was made but never put into full-scale production. It had a small cargo platform over the front wheel, while the operator sat behind it.

Specifications

- Designation: Motocicletta Volugrafo Aermoto
- Producer: Officine Meccaniche Volugrafo, Turin
- Years produced: 1942–1944
- Number produced: 600 (estimated)
- Length: 1,050mm (41.5 inches)
- Width: 310mm (12.2 inches) with handlebars folded
- Height: 530mm (20.87 inches)
- Unladen weight: empty 51.5kg (113.5lbs); equipped 59kg (130lbs)
- Tyres: 2.50 x 8
- Engine: Batua single cylinder, two-stroke, 120cc, 2 HP @3,600 rpm
- Transmission: two speeds, chain-driven
- Fuel capacity: 9.5 litres (2.5 US gallons, 2 Imperial gallons)
- Fuel: gasoline/oil mixture
- Speed: 50km/h (31 mph)
- Range (on road): 300km (186 miles)

Motociclo Benelli 250 M37

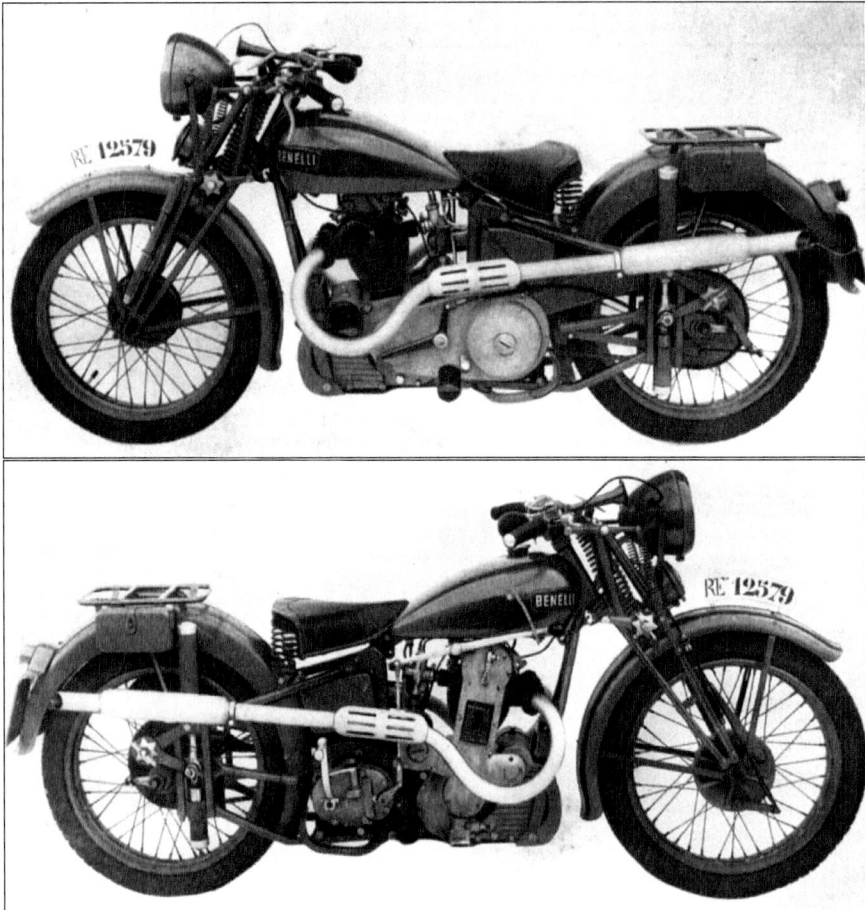

The Benelli 250 M37 was a military model derived from the civilian Benelli 250 TN. Of note are the two silencers, one on each side.

Developmental and Service History

The Benelli 250 M37 TE was a single-seat motorcycle whose production began in 1937; it was one of three types of Benelli motorcycle adopted by the *Regio Esercito* prior to the Second World War with which it shared many components, including engine, transmission, brakes and wheels. Motorcycles with 250cc engines like this were used in the army mainly for messenger and escort duties, leaving vehicles with 500cc engines (see for example the Benelli VL 500) at the disposal of the fighting troops. Details concerning numbers built and specific service history are lacking.

It should be noted that the Benelli 250 and almost all the other motorcycles mentioned in this work were also used by the Royal Carabinieri.

Technical Description

The Benelli 250 M37 TE (TE meant *militare mod. 1937 Telaio Elastico*, or military model 1937 with elastic frame) derived from the civilian 250 TN; it was a conventionally configured motorcycle with a single cylinder four-stroke engine inclined 12 degrees forward, manual transmission, a front fork with compression springs and adjustable dampers and a patented rear swing arm with shock absorbers. The fuel tank was above the engine, and the single seat was directly behind the fuel tank. The transmission was activated by a pedal and consisted of a primary gear and secondary chain. The front brake was activated by hand, the rear brake by a pedal. The electrical system was based on a Marelli dynamo.

Specifications

- Designation: Motociclo Monoposto Benelli 250 M37 TE
- Producer: Fabbrica Motoveicoli Fratelli Benelli, Pesaro
- Years produced: 1937–?
- Number produced: NA
- Length: 2,160mm (85 inches)
- Width: 730mm (28.74 inches)
- Height: 1,000mm (39.37 inches)
- Unladen weight: 147kg (324lbs)
- Unladen weight (with fuel, oil and tools): 161kg (355lbs)
- Wheelbase: 1397mm (55 inches)
- Tyres: Superflex 3.00 x 19
- Minimum turning radius: inclined vehicle 2,000mm (78.74 inches); upright vehicle 2,200mm (86.61 inches)
- Minimum clearance: 150mm (5.9 inches)
- Fording depth: 350mm (13.78 inches)
- Engine: Benelli 4TM single cylinder, four-stroke, air-cooled, 246.79cc, 9 HP @4,750 rpm
- Transmission: four speeds
- Fuel capacity: 12 litres (3.17 US gallons, 2.64 Imperial gallons)
- Speed: 66km/h (41 mph)
- Range (on road): 200km (125 miles)
- Range (cross-country): 5 hours

Motociclo Benelli 500 VL Militare

The Benelli 500 VL Militare, derived from the civilian model 500 VLC.

The Benelli 500 VL Militare, derived from the civilian model 500 VLC.

Developmental and Service history

The Benelli 500 VLM (*valvole laterali militare*, that is side valves, military) aka 500 VL TE M40 (*telaio elastico, militare modello 1940*; elastic frame model military 1940) derived from the civilian model 500 VLC. This motorcycle could be configured with either one or two seats. It shared many mechanical components with the smaller Benelli 250 M37.

Similarly to the Guzzi Alce 500, in the army the 500 VLM was employed for reconnaissance and liaison duties, for example equipping entire Bersaglieri (mobile light infantry) motorcycle battalions of the armoured and motorised divisions.

Details concerning numbers built and specific service history are lacking.

Technical Description

The Benelli 500 VLM was a conventionally configured motorcycle with a single cylinder four-stroke engine inclined 3 degrees forward, manual transmission, a front fork with compression springs and adjustable dampers and a patented rear swing arm with shock absorbers. The fuel tank was above the engine, and the seat of the single-seat version (*monoposto*) was directly behind the fuel tank, while that of the two-seat version (*biposto*) was over the rear fender. The transmission was activated by a pedal and consisted of a primary gear and secondary chain. The front brake was activated by hand, the rear brake by a pedal. The electrical system was based on a Marelli D30 dynamo, without battery.

Specifications

- Designation: Motociclo Militare Benelli 500 VL TE M40
- Producer: Fabbrica Motoveicoli Fratelli Benelli, Pesaro
- Years produced: 1940–?
- Number produced: Not known
- Length: 2,130mm (83.8 inches)
- Width: 800mm (31.5 inches)
- Height: 1,020mm (40.16 inches)
- Unladen weight: 178kg (392lbs) *monoposto*; 187kg (412lbs) *biposto*
- Wheelbase: 1,400mm (55.1 inches)
- Tyres: Superflex 3.50 x 19
- Minimum turning radius: 2,300 – 2,500mm (90.5 – 98.5 inches)
- Minimum clearance: 200mm (7.87 inches)
- Engine: Benelli single cylinder, four-stroke, air-cooled, 493.6cc, 11 HP @4,200 rpm
- Transmission: four speeds
- Fuel capacity: 12.5 litres (3.30 US gallons, 2.75 Imperial gallons)
- Speed: 86.5km/h (53.7 mph)
- Range (on road): 275km (171 miles) monoposto; 250km (155 miles) biposto

Mototriciclo Benelli 500 M36

The Benelli 500 M36 motor tricycle was the three-wheeled adaptation of the Benelli 500 VL motorcycle.

In view of the attack on France in 1940, a formation of Benelli 500 M36 equipped with a radio set R2 mod. 1935 (a radiotelegraphic station for infantry troops with an range of about 10km) was organised. (ACS)

A column of Benelli 500 M36 carrying cargo as well as personnel (Bersaglieri). Flanking the head of the column of trucks is a Fiat 1100 Camioncino. (V. de Gaetano)

Developmental and Service History

The mototriciclo (motor tricycle) Benelli 500 M36 was the three-wheeled adaptation of the Benelli 500 VL motorcycle. Details concerning numbers built are missing.

The mototriciclo Benelli was requested in the mid-thirties by the *Regio Esercito* to motorise some infantry units, in particular the Bersaglieri who were considered 'rapid troops'. It began to be used in the Spanish Civil War, both for the transport of ammunition, light weapons and other materials, and the towing of 20mm Breda automatic cannons. Later, it was joined in all these roles by the Guzzi Trialce (see below).

Technical Description

The mototriciclo Benelli 500 was a three-wheeled motorcycle that shared many mechanical components of the Benelli 500 VL from which it derived. It had a single cylinder, four-stroke engine inclined 3 degrees forward, manual transmission and a front fork with compression springs and adjustable dampers; the rear swing arm and single wheel were replaced by a steel frame that held the two rear wheels with a suspended axle on leaf springs. The frame also held a wooden box body that had a 600kg carrying capacity. The fuel tank was above the engine, and the seat was directly behind the fuel tank. The transmission was activated by a pedal and consisted of a primary gear and secondary chain, acting on the rear differential gear. The front brake was activated by hand, the rear brakes by a pedal. The electrical system was based on a Marelli dynamo.

Specifications

- Designation: Mototriciclo Benelli 500 M36
- Producer: Fabbrica Motoveicoli Fratelli Benelli, Pesaro
- Years produced: 1936–?
- Number produced: NA
- Length: 3,020mm (118.9 inches)
- Width: 1,250mm (49.2 inches)
- Height: 1,020mm (40.16 inches)
- Unladen weight: 327kg without liquids (720.9lbs); 346kg (762.8lbs) in travelling order
- Wheelbase: 1,924mm (75.74 inches)
- Rear track: 900mm (35.43 inches)
- Tyres: Superflex 3.50 x 19 (old nomenclature 26 x 3.50)
- Minimum turning radius: 2,000 – 2,300mm (79 – 90.5 inches)
- Minimum ground clearance: 250mm (9.84 inches)
- Carrying capacity: 370kg (815lbs)
- Box body length: 1,240mm (48.82 inches)
- Box body width: 900mm (35.43 inches)
- Box body sides height: 320mm (12.6 inches)
- Engine: Benelli 4 TMN 500, single cylinder, four-stroke, air-cooled, 493.6cc, 12HP @4,500rpm
- Transmission: four speeds
- Fuel capacity: 13.5 litres (3.5 US gallons, 3 Imperial gallons)
- Speed: 72.6km/h (45 mph)
- Range (on road): 225 – 260km (140 – 162 miles)

Motociclo Bianchi 500 M

The Bianchi 500 M single-seat version (*monoposto*).

The Bianchi 500 M single-seat version (*monoposto*).

The Bianchi 500 M two-seat version (*biposto*).

Bersaglieri and a Bianchi 500 M with leg shields in North Africa in the spring of 1941. (BAMA)

A Bianchi 500 M single seat with standard Pirelli Superflex tyres. Note the *Regio Esercito* badge under the headlight.

The Bianchi motorcycle also operated on the Eastern front; shown here in 1942 with some Bianchi Miles trucks under repair. (L. Valente)

Developmental and Service History

In addition to supplying staff cars and trucks for Italian military forces during the Second World War, the Edoardo Bianchi company also supplied motorcycles that saw service in Italian East Africa and later in Spain. Beginning in 1936, Bianchi began to produce a new 500cc military motorcycle, the 500 M (*militare*, military) which operated throughout the war in Italy and in North Africa. The 500 M was produced in both a single-seat as well as a two-seat version. The Bianchi 500 M continued to be used by the Italian Army into the 1950s.

Technical Description

The Bianchi 500 M was a conventionally configured motorcycle with a single cylinder four-stroke engine mounted vertically, manual transmission, a front fork with compression springs and a rear swing arm. The fuel tank was above the engine, and the seat of the single seat version (*monoposto*) was directly behind the fuel tank, while that of the two-seat version (*biposto*) was over the rear fender. The transmission was activated by a pedal. The front brake was activated by hand, the rear brake by a pedal. The 6-volt electrical system used a Marelli D30 dynamo.

Specifications

- Designation: Motociclo Bianchi 500 M
- Producer: Edoardo Bianchi, Milan
- Years produced: 1936–1944
- Number produced: Not Known
- Length: 2,120mm (83.46 inches)
- Width: 750mm (29.5 inches)
- Height: 960mm (37.8 inches)
- Unladen weight: 170kg (375lbs) *monoposto*; 178kg (392lbs) *biposto*
- Wheelbase: 1,380mm (54.33 inches)
- Tyres: Superflex 3.50 x 19
- Minimum turning radius: 2,300mm (90.55 inches)
- Minimum clearance: 165mm (6.5 inches)
- Engine: Bianchi single cylinder, four-stroke, air-cooled, 498cc, 9 HP @3,200 rpm
- Transmission: three speeds
- Fuel capacity: 12 litres (3.2 US gallons, 2.6 Imperial gallons)
- Speed: 75km/h (47mph)
- Range (on road): 260km (162 miles) *monoposto*; 240km (149 miles) *biposto*
- Range (off-road): 200km (124 miles) *monoposto*

Motociclo Gilera 500 LTE

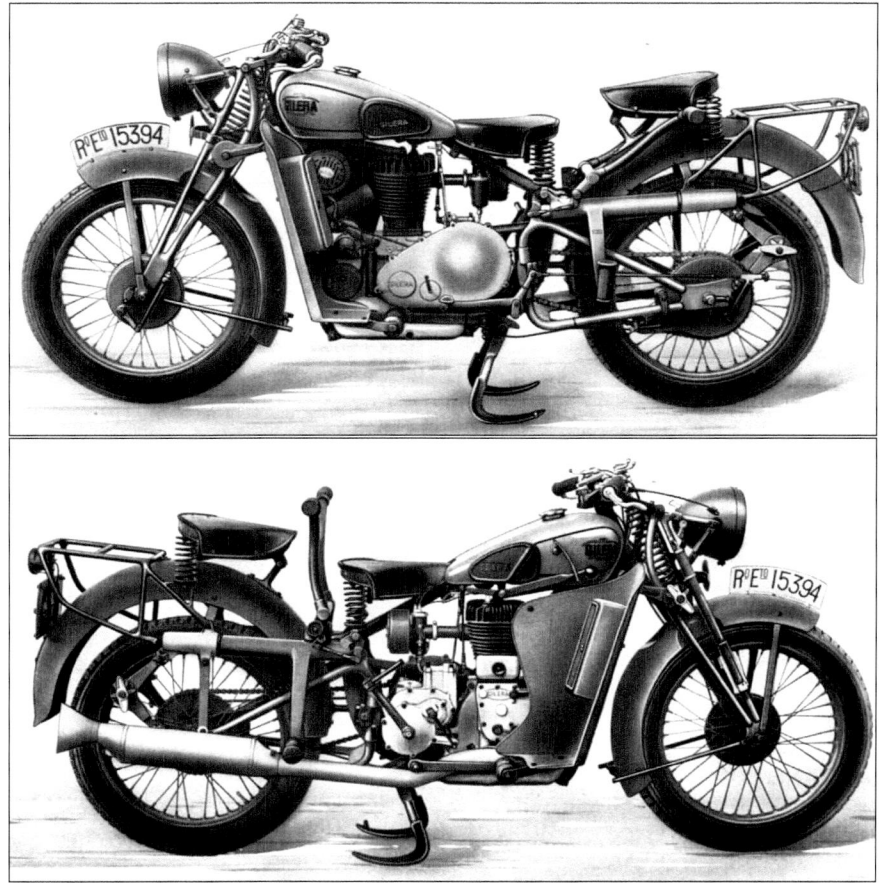

The Gilera 500 LTE.

Gilera 500 LTE with Breda 30 light machine gun.

An Italian column on the move in Egyptian territory. In the foreground, a Gilera 500 LTE in front of two Fiat 508 CMs. (ACS)

Spring 1941. Bersaglieri from the 7° *Reggimento* in the Tobruk area. (ACS)

Developmental and Service History

Gilera started its own production of motorcycles in 1909; its products had a reputation for having exceptional performance. In 1937 the company began manufacturing a 500cc motorcycle for the *Regio Esercito*. It was produced in single-seat and two-seat configurations as well as in a three-wheeled version. The 500 LTE (originally VLTE, *Valvole Laterali, Telaio Elastico*, that is side valves, Elastic Frame) served on all fronts during the Second World War and remained in production until 1944, as well as serving in the post-war Italian Army for a while. Production figures are unknown.

Technical Description

The Gilera 500 LTE was a motorcycle powered by a vertically mounted single cylinder four-stroke engine. The tubular steel frame consisted of a parallelogram front fork and a patented flexible rear arm. The four-speed transmission was operated by a lever to the right of the gas tank; the LTE had a chain drive and had drum brakes. The fuel tank was above the engine, and the seat of the single seat version (*monoposto*) was directly behind the fuel tank, while that of the two-seat version (*biposto*) was over the rear fender; there was a set of dummy handlebars between the front and rear seats for the passenger to hold on to. The 6-volt electrical system was based on a Marelli D30 dynamo.

Specifications

- Designation: Motociclo Gilera 500 LTE Militare
- Producer: Gilera, Arcore
- Years produced: 1937–1944
- Number produced: Not Known
- Length: 2,200mm (86.61 inches) *monoposto*; 2,260mm (88.98 inches) *biposto*
- Width: 800mm (31.5 inches)
- Height: 1,050mm (41.34 inches)
- Unladen weight: 190kg (419lbs) *monoposto*; 203kg (448lbs) *biposto*
- Wheelbase: 1,450mm (57 inches) *monoposto*; 1,500mm (59 inches) *biposto*
- Tyres: Superflex 3.50 x 19
- Minimum turning radius: inclined vehicle 1,400mm (55.12 inches); upright vehicle 1,600mm (63 inches)
- Minimum clearance: 140mm (5.51 inches)
- Engine: Gilera 500 L single cylinder, four-stroke, air-cooled, 498cc, 12HP @3,800 rpm
- Transmission: four speeds
- Fuel capacity: 12 litres (3.2 US gallons, 2.6 Imperial gallons)
- Speed: 76km/h (47mph), later 80km/h (49mph) *monoposto*; 80km/h (49mph) *biposto*
- Range (on road): 230km (143 miles) *monoposto*; 220km (137 miles) *biposto*

Motocarrozzetta Gilera Marte 500

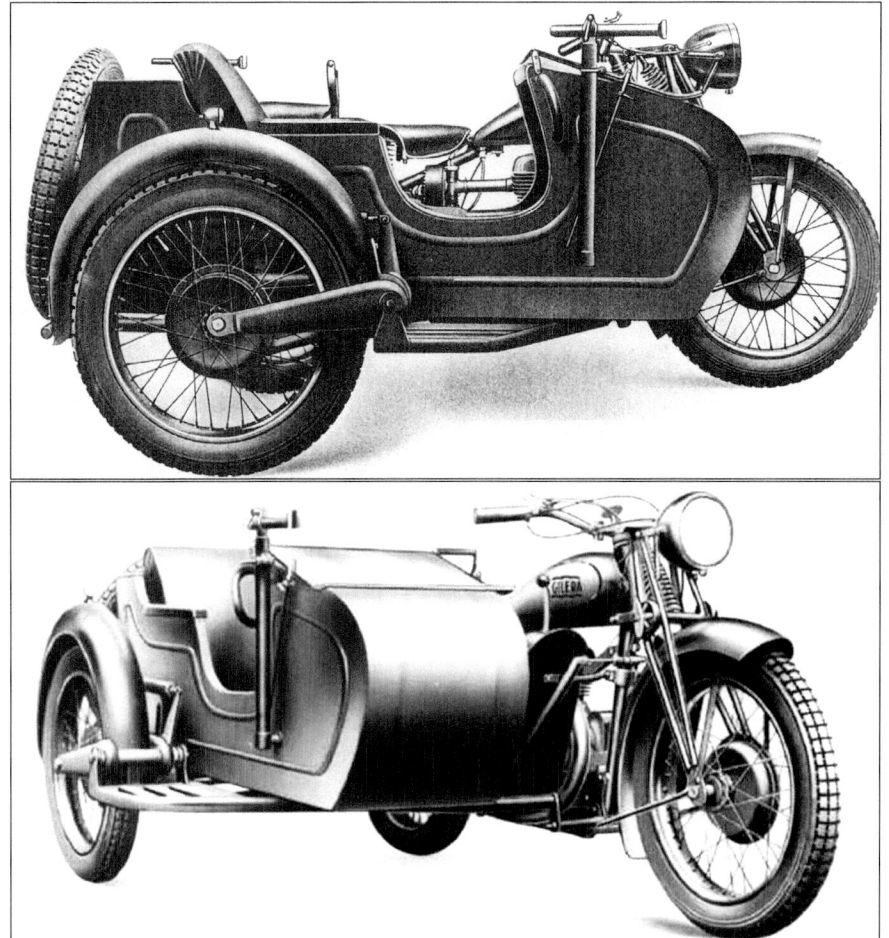

The Gilera Marte 500 sidecar.

Rear view of the Marte 500.

Developmental and Service History

At the end of 1940, Gilera designed the Marte 500, referred to as a *motocarrozzetta* (sidecar) based on the Gilera LTE 500 motorcycle (see above). The *Marte* (Mars) is said to have been inspired by the German sidecar configuration. It was developed to replace motorcycles equipped with light machine guns mounted on the handlebars in the motorcycle platoons and served throughout the war. After the Armistice it was tested by the *Oberkommando Wehrmacht*, which highlighted its defects, namely low power and high weight. At the end of the war, 158 that were completed and stored in a warehouse were converted to a civilian configuration. Total production figures are unavailable.

Technical Description

The Gilera Marte 500 was a motorcycle powered by a vertically mounted single cylinder four-stroke engine. A Cardan drive transmitted power to the rear wheel of the motorcycle, while a second shaft to the sidecar's wheel could be disengaged while the Marte was travelling on the road. The sidecar's wheel could oscillate freely on its own fork. The sidecar body had a mount on which a Breda model 30 light machine gun was fixed; a spare wheel with its tyre was mounted on the rear of the sidecar.

Specifications

- Designation: Motocarrozzetta Gilera Marte 500
- Producer: Gilera, Arcore
- Years produced: 1941–1945
- Number produced: Not Known
- Length: 2,300mm (90.55 inches)
- Width: 1,600mm (63 inches)
- Height: 1,020mm (40.15 inches)
- Unladen weight: 300kg (661lbs)
- Wheelbase: 1,400mm (55.11 inches)
- Rear track: 1,500mm (59 inches)
- Tyres: Superflex 3.50 x 19
- Minimum clearance: 180mm (7 inches)
- Minimum turning radius: 5,000mm (197 inches)
- Carrying capacity: 240kg (529lbs)
- Engine: Gilera 500 L single cylinder, four-stroke, air-cooled, 498cc, 14 HP @4,800 rpm
- Transmission: four speeds
- Fuel capacity: 16 litres (4.2 US gallons, 3.5 Imperial gallons)
- Speed: 78km/h (48 mph)
- Range (on road): 220km (137 miles)
- Range (cross-country): 180km (112 miles)

Motocarro Gilera Mercurio

The Gilera Mercurio motor tricycle with tarpaulin cover. Drawing from the user manual.

The Gilera Mercurio without tarpaulin.

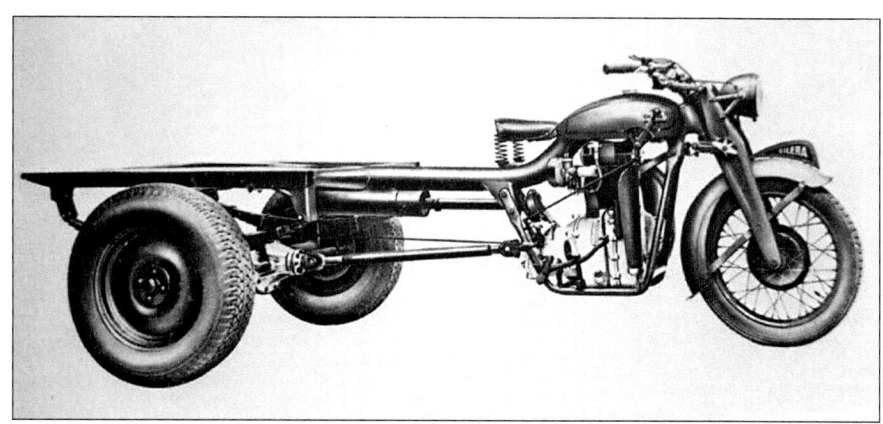

The frame of the Mercurio.

Developmental and Service History
The Gilera Mercurio was a three-wheeled motorcycle designed as *motocarro* (three-wheeler light truck) and based on the Gilera 500 LTE motorcycle. Gilera developed the *Mercurio* (Mercury) in 1938, but production was delayed until 1940. The Mercurio served on all fronts during the war. The September 1943 Armistice put an end to production; some or most of the Mercurios in the *Regio Esercito* inventory were then appropriated by Mussolini's *Esercito Nazionale Repubblicano* (National Republican Army) in northern Italy. After the war, production of a civilian version of the Mercurio continued until 1963.

Technical Description
The Gilera Mercurio shared the same mechanical components as the 500 LTE motorcycle from which it derived (see above). It had a single cylinder four-stroke engine; its transmission had four speeds forward as well as one reverse speed; power to the rear wheels was by a Cardan drive. The single rear wheel was replaced by a set of two wheels mounted on a frame behind the driver's seat; the frame held a small wooden cargo body with raised sides and tailgate which could be covered by a canvas cover supported by three bows. The normal motorcycle wire-spoke wheels were replaced by all-metal pressed steel wheels in the rear, while the wire-spoke wheel remained as the front wheel. The Mercurio had drum brakes; the rear brakes could be locked to stop the vehicle on sloping roads. The 6 V electrical system used a Marelli D30 dynamo.

Specifications
- Designation: Motocarro Gilera Mercurio
- Producer: Gilera, Arcore
- Years produced: 1940–1963
- Number produced: approximately 1,000 produced during the Second World War; production of a civilian version continued until 1963
- Length: 3,560mm (140.15 inches) standard type (short); 1,600mm (148.03 inches) lengthened type
- Width: 1,500mm (59 inches) standard type (short); 1,600mm (62.99 inches) lengthened type
- Height: 1,400mm (55.11 inches)
- Unladen weight: 580kg (1,278lbs) standard type (short); 610kg (1,385lbs) lengthened type
- Wheelbase: 2,230mm (87.8 inches) standard type (short); 2,430mm (95.67 inches) lengthened type
- Rear track: 1,260mm (49.6 inches)
- Tyres: Superflex 3.50 x 19 front; 6.00 x 16 rear
- Minimum clearance: 195mm (7.68 inches)

- Minimum turning radius: 3,900mm (154 inches)
- Carrying capacity: 1,500kg (3,307lbs); 1,000kg (2,204lbs) standard type.
- Box body length: 2,020mm (79.52 inches) standard type (short); 2,430mm (95.67 inches) lengthened type
- Box body width: 1,420mm (55.9 inches) standard type (short); 1,550mm (61.02 inches) lengthened type
- Box body sides height: 390mm (15.35 inches) standard type; 405mm (15.95 inches) lengthened type
- Engine: Gilera 500 L single cylinder, four-stroke, air-cooled, 498cc, 18 HP @4,100 rpm
- Transmission: four speeds forward, one reverse
- Fuel capacity: 15 litres (4 US gallons, 3.3 Imperial gallons)
- Speed: 70km/h (43mph)
- Range (on road): 200km (124 miles)

Motociclo Sertum 500 MCM

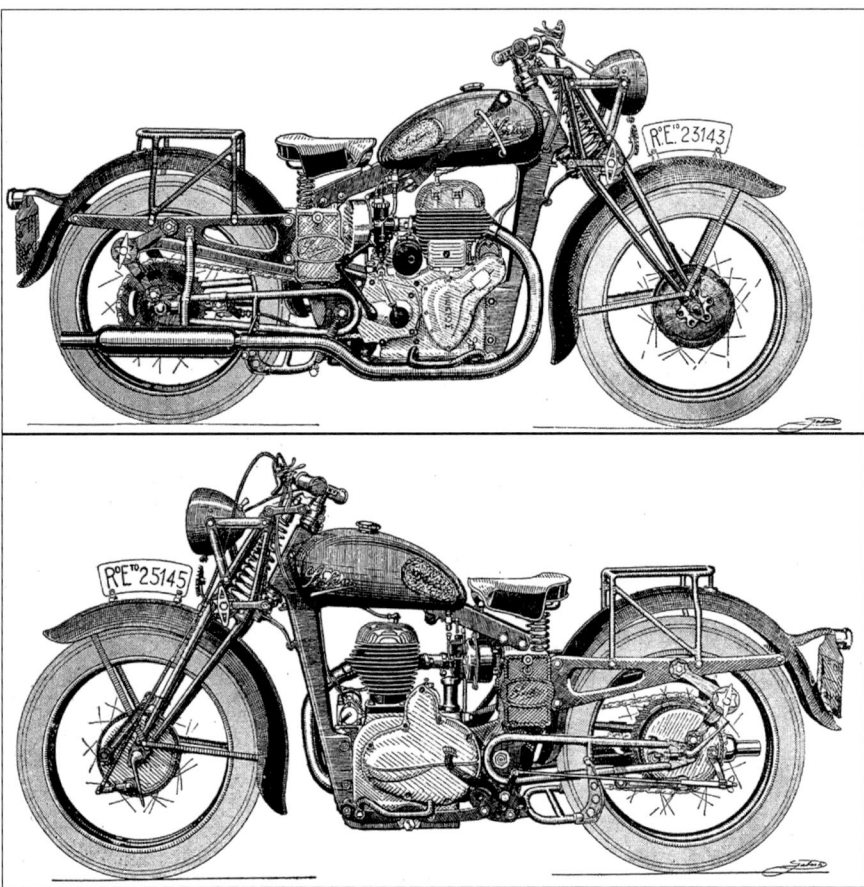

The Sertum 500 MCM single-seat.

The Sertum 500 MCM two-seat.

The Sertum 500 MCM two-seat.

The three-wheeled Motocarro 500 MCM.

A Sertum 500 MCM single-seat with *Regio Esercito* number plate. (ACS)

Developmental and Service History

The firm of Fausto Alberti of Milan, specialised in industrial engines and mechanical parts, was a relative newcomer to the motorcycle field, not having been established until 1932, marketing its machines under the brand name of Sertum (the Latin word for wreath). By 1937 Sertum had achieved a status equal to that of the more established names of Benelli, Bianchi, Gilera and Moto Guzzi. With the advent of the Second World War, Sertum developed a 500cc military machine in both single-seat (*monoposto*) and two-seat (*biposto*) versions, unveiling its 500 MCM in 1941. The vehicle derived from the civilian version 500

VL (*Valvole Laterali*). Subjected to rigorous testing, it was received enthusiastically by the military authorities. It was considered to be the most modern motorcycle assigned to motorcycle units in the Greek and North African campaigns. It remained in service with the Italian Army until the mid–1950s.

Technical Description

The Sertum 500 MCM was a conventionally configured motorcycle with a single cylinder four-stroke engine with side valves mounted vertically. The suspension's front fork used central compression coil springs; the rear suspension consisted of a swing arm with a leaf spring located under the seat. The fuel tank was above the engine, and the seat of the single seat version was directly behind the fuel tank, while that of the two-seat version was over the rear fender; there was a set of foldable handlebars between the front and rear seats for the passenger to hold on to. The transmission was activated by a pedal. The front drum brake was activated by hand, the rear drum brake by a pedal. The 6 V electrical system used a Marelli D30 dynamo.

Variants

A three-wheeled version, the Motocarro 500 MCM, was produced. It was similar to the Mototriciclo Benelli 500 M36 and the Guzzi Trialce (see separate entries). Its wooden box body measured externally 2,020mm (6ft 7in) x 1,420mm (4ft 8in) x 350mm (1ft 2in); it had a 500kg (1,102lbs) maximum carrying capacity.

Specifications

- Designation: Motociclo Sertum 500 MCM
- Producer: Officine Meccaniche Fausto Alberti, Milan
- Years produced: 1941–1946
- Number produced: Not Known
- Length: 2,190mm (86.22 inches)
- Width: 850mm (33.46 inches)
- Height: 1,070mm (42.12 inches)
- Unladen weight: 104kg (229lbs) *monoposto*; 113kg (249lbs) *biposto*
- Unladen weight (with fuel, oil and tools): 177kg (390lbs) *monoposto*; 186kg (410lbs) *biposto*
- Wheelbase: 1,450mm (57 inches)
- Tyres: Superflex 3.50 x 19
- Minimum clearance: 170mm (6.69 inches)
- Engine: Alberti Tipo 500 MCM single cylinder, four-stroke, air-cooled, 498cc, 12 HP @4,100 rpm
- Transmission: four speeds
- Fuel capacity: 13 litres (3.43 US gallons, 2.85 Imperial gallons)
- Speed: 77km/h (48 mph) *monoposto*; 73km/h (45 mph) *biposto*
- Range (on road): 200km (125 miles)

Motociclo Guzzi 500 Alce

The Guzzi 500 Alce single-seat. Note the characteristic double silencer on the left side. (Moto Guzzi)

The Guzzi 500 Alce two-seat. (Moto Guzzi)

A *biposto* (two-seat) seen from the right side. (Moto Guzzi)

The sidecar version. (Moto Guzzi)

Developmental and Service History

The Guzzi 500 *Alce* (elk, moose) stemmed from a number of earlier civilian Guzzi motorcycles since 1928, beginning with the GT, then followed by the Sport 14 in 1930, then by the GT 17 in 1932, GTV in 1934, S in 1937, and in 1939 by the transitional GT 20. The Guzzi 500 Alce – as well as the previous models – was adopted by the *Regio Esercito* and, starting from 1940, was widely issued to the three Italian armed forces and to the *Polizia dell'Africa Italiana*. In the army it was assigned to each infantry regiment for reconnaissance and liaison duties, and in the armoured and motorised divisions in North Africa it equipped entire Bersaglieri motorcycle battalions.

The Alce was issued in three versions: a single-seat (*monoposto*), a two-seat (*biposto*), and a version with a removable sidecar (*motocarrozzetta*). Some Alces were fitted with a special support for a 6.5mm Breda model 30 light machine gun; the gun could

The heavy machine gun carrier version. (Moto Guzzi)

Bersaglieri, riding Guzzi motorcycles, about to cross a rough wooden bridge as Rumanian locals look on in the background. (AUSSME)

not be used while the machine was moving, but the support enabled it to be used while stationary. Another variant included the 8mm Breda model 37 machine gun, which was heavier than the model 30, so it could only be carried hanging on the left side, while the tripod was hinged on the right side. Production of the Alce continued throughout the war until 1945.

Technical Description

The Guzzi 500 Alce was a conventionally configured motorcycle which had a horizontal four-stroke engine, manual transmission, a parallelogram front fork with friction dampers and a rear fork likewise with friction dampers; dampers were made by means of discs of friction material (such as cork) tightened between metal discs clamped together. The fuel tank was above the engine, and the seat of the single-seat version was directly behind the fuel tank; the second seat of the two-seat version was slightly raised above the rear fender sand had a separate set of foldable handlebars for the passenger to hold onto. The transmission consisted of a primary gear and secondary chain. The front brake was activated by hand, the rear brake by a pedal. The rear wheel had a detachable hub, i.e. brake and crown were placed on the same side, so that the disassembly of the chain to remove the wheel was

Bersaglieri intent on overcoming a muddy stretch with an Alce. (ACS)

A column of Guzzi GT 17 two-seat, the 'grandmother' of the Guzzi Alce. (AUSSME)

A column of Guzzi motorcycles carrying Bersaglieri in a Ukrainian village; the lead is a Guzzi GT 17 *monoposto* and is fitted with a Breda Model 1930 light machine gun, while most of the other Guzzis in the group are the GT 17 biposto version. (ACS)

not necessary; the operation could be carried out with the motorcycle hoisted on the central stand. Like almost all motorcycles of the time, it was equipped with a 6 V electrical system with Marelli D30 dynamo. It could be equipped with leg guards.

Variants
1,741 examples of a three-wheeled version, the Guzzi Trialce, were also produced (see below).

Specifications
- Designation: Motociclo Guzzi Alce
- Producer: Moto Guzzi, Mandello del Lario
- Years produced: 1940–1945
- Number produced: 6,390 plus 669 with sidecar
- Length: 2,220mm (87.4 inches)
- Width: 790mm (31.1 inches) *monoposto* and *biposto*; 1,575mm (62 inches) sidecar
- Height: 1,065mm (41.93 inches)
- Unladen weight (with fuel, oil and tools): 179.5kg (398lbs) *monoposto*, 187kg (412lbs) *biposto*; 260kg (573lbs) sidecar
- Wheelbase: 1,455mm (57.28 inches)
- Tyres: Superflex 3.50 x 19

Moto Guzzi GT 17s with Breda 30 machine gun.

Another Moto Guzzi GT 17 with Breda 30 machine gun.

- Minimum clearance: 210mm (8.26 inches)
- Engine: Guzzi single cylinder air-cooled, 498.4cc, 13.2 HP @4,000 rpm
- Transmission: four speeds
- Fuel capacity: 13.5 litres (3.5 US gallons, 3 Imperial gallons)
- Speed: 90km/h (56mph) *monoposto*; 85km/h (52mph) *biposto*
- Range (on road): 300km (186 miles)

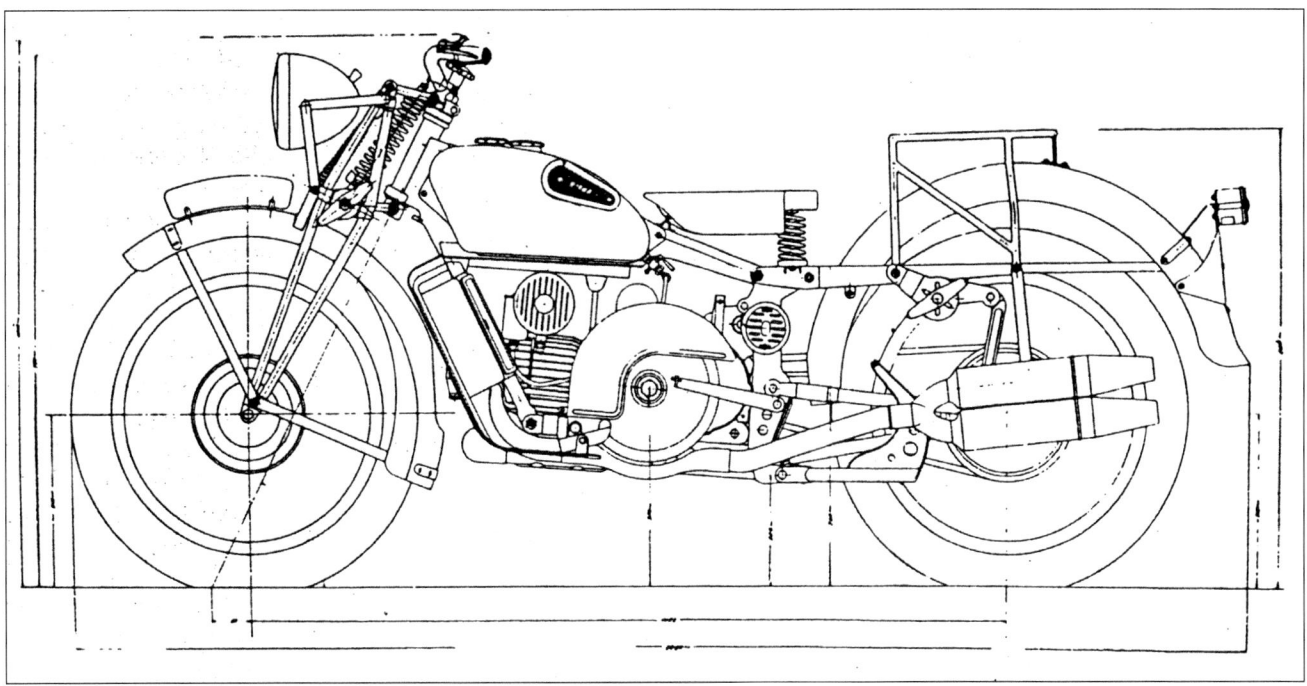

Line drawing of the Alce *monoposto* (left side view). (Guzzi)

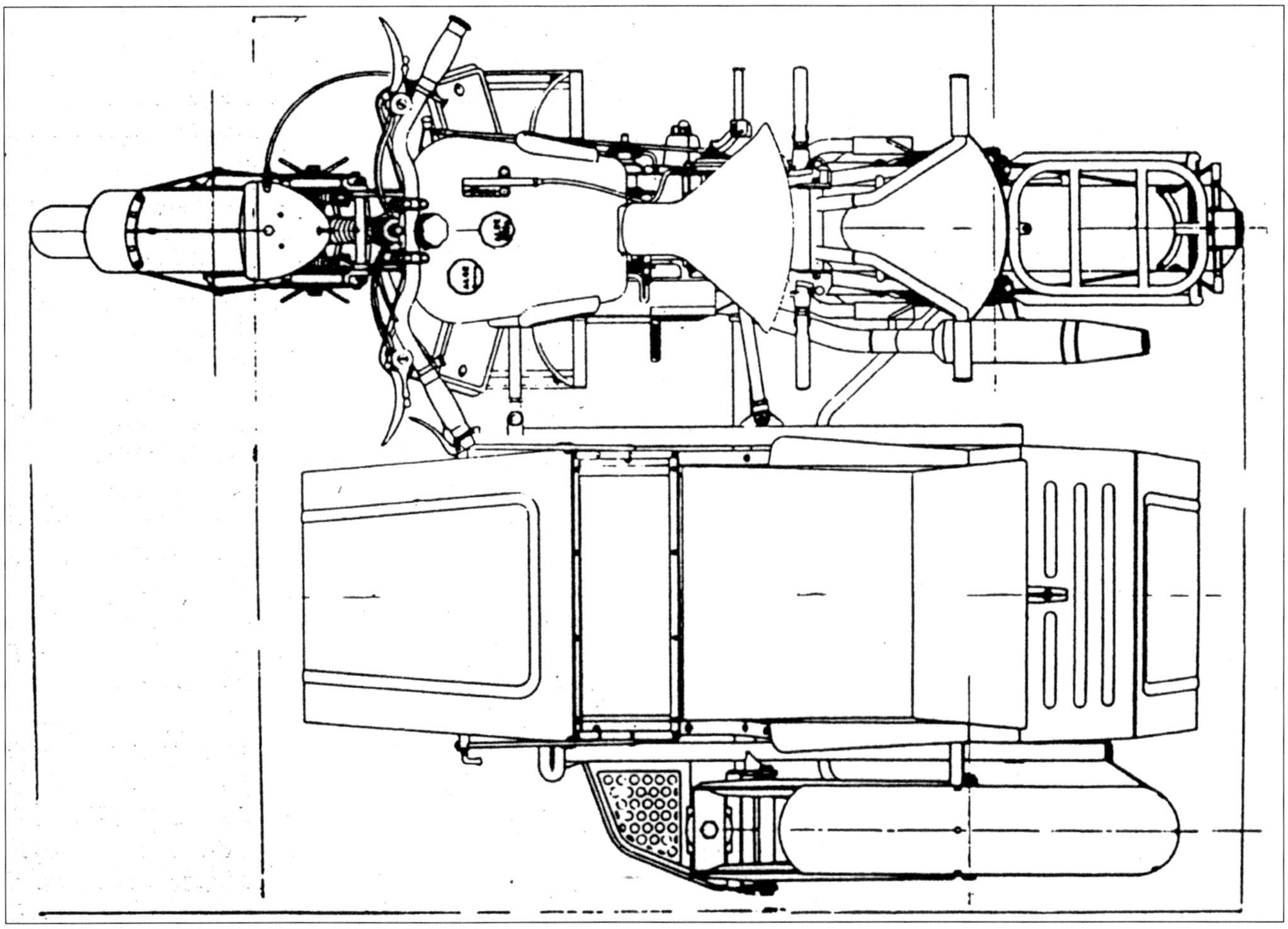

Line drawing (top view) of the Alce Motocarrozzetta. (Guzzi)

Mototriciclo Guzzi Trialce

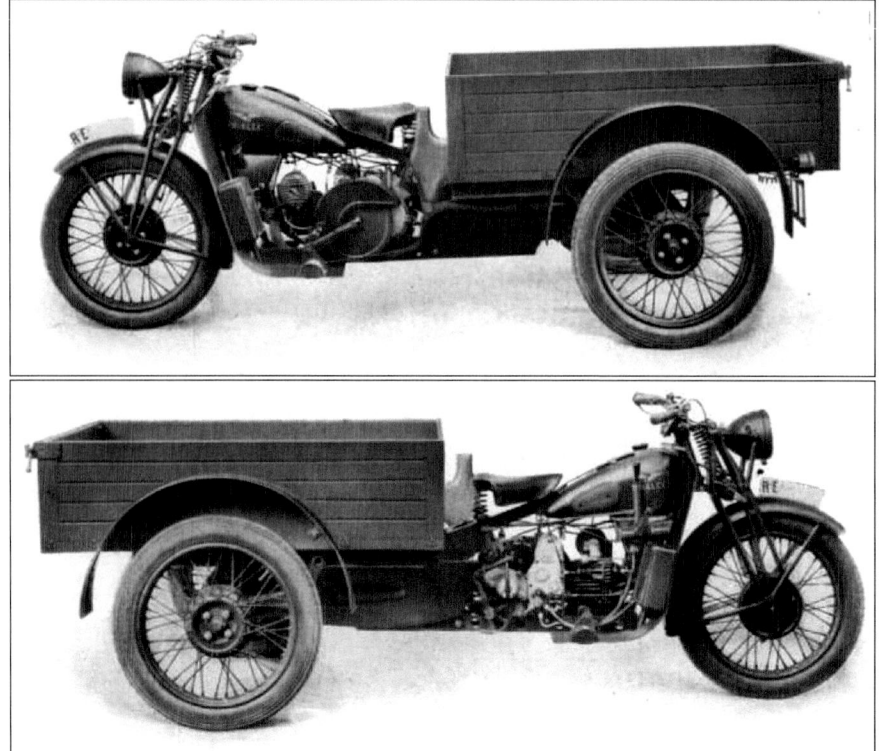

The Guzzi Trialce motor tricycle. (Moto Guzzi)

Front and rear view of the Trialce. (Moto Guzzi)

Developmental and Service History

During the thirties, the Guzzi firm had entered the market of special three-wheeled motorcycles for the transport of materials, called *mototricicli* (motor tricycles). In 1940 the most modern model was launched, called Trialce and based on the proven chassis of the 500 Alce. 1,741 examples were produced until production ceased in 1943.

Beyond general uses, from the end of 1941 the Guzzi Trialce equipped the so-called 'motorcycle-machine-gun companies', who were also equipped with single-seat and two-seat motorbikes as well as trucks. Each Trialce carried, in addition to the machine gun, the gunner, the assistant and the ammunition boxes.

A Bersagliere riding a Guzzi Trialce past the Arco dei Fileni (Arch of the Philaeni), which marked the boundary between the regions of Tripolitania and Cyrenaica in Libya.

Guzzi motor tricycles being used to tow 65/17 guns of the 'La Spezia' Division's *80° Reggimento Artiglieria* in Tunisia, early 1943. (E. Finazzer)

A Guzzi Trialce in the Ukraine being used to transport Russian prisoners. (AUSSME)

The Mototriciclo Guzzi modello 32, manufactured between 1932 and 1939, was the predecessor of the Trialce. Photographed here in use in North Africa.

Technical Description

The Trialce used the front chassis, the fork and the engine of the Alce, a single cylinder four-stroke. The primary transmission was made of helical gears, the secondary used a roller chain, with a four-speed gearbox. The rear of the structure was modified with the installation of a suspension axle frame on helical springs and chain drive on the central differential. The frame supported a wooden bed. The wheels were interchangeable. The brakes remained drum brakes, the front ones operated by hand and the rear ones by pedal.

Another ancestor of the Trialce was the Guzzi ER manufactured from 1938. Here an example belonging to the *Unione Nazionale Protezione Antiaerea* (National Organisation for Anti-aircraft Protection); UNPA employed civilian personnel, who were mostly female.

Nice close-up of a captured Trialce. The writing relating to tyre inflation pressure stands out against the grey-green colour.

Variants

A variant of the Trialce could carry, in addition to the personnel, a Breda 37 machine gun on an anti-aircraft carriage, plus a tripod for ground fire.

The most original version was undoubtedly the air-transportable version assigned to the paratroops that could be disassembled.

Specifications

- Designation: Mototriciclo Guzzi Alce
- Producer: Moto Guzzi, Mandello del Lario
- Years produced: 1940–1943
- Number produced: 1,741
- Length: 2,825mm (9ft 3in)
- Width: 1,240mm (4ft 1in)
- Height: 1,050mm (3ft 5in)
- Unladen weight (with fuel, oil and tools): 354kg (780lbs)
- Wheelbase: 1,880mm (6ft 2in)
- Tyres: Superflex 3.50 x 19
- Minimum clearance: 210mm (8.26 inches)
- Carrying capacity: 400kg (881.85lbs) normal; 500kg (1,102lbs) maximum
- Box body length: 1,300mm (4ft 3in) external; 1,250mm (4ft 2in) internal
- Box body width: 950mm (3ft 1in) external; 900mm (2ft 11in) internal
- Box body sides height: 350mm (1ft 1in) external; 315mm (1ft) internal
- Engine: Guzzi single cylinder air-cooled, 498.4cc, 13.2 HP @4,000 rpm
- Transmission: four speeds
- Fuel capacity: 16 litres (4.2 US gallons, 3.5 Imperial gallons)
- Speed: 73km/h (45mph)
- Range (on road): 260km (161 miles)

A Bersagliere riding a Guzzi Trialce past the Arco dei Fileni (Arch of the Philaeni), which marked the boundary between the regions of Tripolitania and Cyrenaica in Libya.

Guzzi motor tricycles being used to tow 65/17 guns of the 'La Spezia' Division's *80° Reggimento Artiglieria* in Tunisia, early 1943. (E. Finazzer)

A Guzzi Trialce in the Ukraine being used to transport Russian prisoners. (AUSSME)

The Mototriciclo Guzzi modello 32, manufactured between 1932 and 1939, was the predecessor of the Trialce. Photographed here in use in North Africa.

Technical Description

The Trialce used the front chassis, the fork and the engine of the Alce, a single cylinder four-stroke. The primary transmission was made of helical gears, the secondary used a roller chain, with a four-speed gearbox. The rear of the structure was modified with the installation of a suspension axle frame on helical springs and chain drive on the central differential. The frame supported a wooden bed. The wheels were interchangeable. The brakes remained drum brakes, the front ones operated by hand and the rear ones by pedal.

Another ancestor of the Trialce was the Guzzi ER manufactured from 1938. Here an example belonging to the *Unione Nazionale Protezione Antiaerea* (National Organisation for Anti-aircraft Protection); UNPA employed civilian personnel, who were mostly female.

Nice close-up of a captured Trialce. The writing relating to tyre inflation pressure stands out against the grey-green colour.

Variants

A variant of the Trialce could carry, in addition to the personnel, a Breda 37 machine gun on an anti-aircraft carriage, plus a tripod for ground fire.

The most original version was undoubtedly the air-transportable version assigned to the paratroops that could be disassembled.

Specifications

- Designation: Mototriciclo Guzzi Alce
- Producer: Moto Guzzi, Mandello del Lario
- Years produced: 1940–1943
- Number produced: 1,741
- Length: 2,825mm (9ft 3in)
- Width: 1,240mm (4ft 1in)
- Height: 1,050mm (3ft 5in)
- Unladen weight (with fuel, oil and tools): 354kg (780lbs)
- Wheelbase: 1,880mm (6ft 2in)
- Tyres: Superflex 3.50 x 19
- Minimum clearance: 210mm (8.26 inches)
- Carrying capacity: 400kg (881.85lbs) normal; 500kg (1,102lbs) maximum
- Box body length: 1,300mm (4ft 3in) external; 1,250mm (4ft 2in) internal
- Box body width: 950mm (3ft 1in) external; 900mm (2ft 11in) internal
- Box body sides height: 350mm (1ft 1in) external; 315mm (1ft) internal
- Engine: Guzzi single cylinder air-cooled, 498.4cc, 13.2 HP @4,000 rpm
- Transmission: four speeds
- Fuel capacity: 16 litres (4.2 US gallons, 3.5 Imperial gallons)
- Speed: 73km/h (45mph)
- Range (on road): 260km (161 miles)

3

Motor Cars

Autovettura Alfa Romeo 2500 C

The civilian Alfa Romeo 2500 6C modified for testing in East Africa. The tools behind the trunk and the additional fuel tanks located at the base of the rear mudguards and next to the spare wheels can all be clearly seen.

A 2500 in Eritrea in 1939, more precisely in Asmara and the Cheren (Keren) area. It bears the regulation 'Prova' number plate for vehicles not yet registered to be subjected to tests and trials. Tyres are Aerflex Stella Bianca. (G.B. Guidotti)

Developmental and Service History

In 1938 the Minister of War set forth a request for a colonial version of the Alfa Romeo 2500 6C (Turismo and Sport versions), to be used as staff car for higher-ranking officers and national authorities. Two examples were ready by 1939, and in October of that year began field trials in Eritrea and Ethiopia. The cars were subjected to a great deal of cross-country travel, as well as periods of travel on improved roads. In particular, the engineers wanted to find a correct setting of the carburettor at high altitudes and to check the oil cooling system. The first ever use of the cars took place in the Addis Ababa area (elevation over 2,400 m).

A 2500 in Eritrea in 1939, more precisely in Asmara and the Cheren (Keren) area. It bears the regulation 'Prova' number plate for vehicles not yet registered to be subjected to tests and trials. Tyres are Aerflex Stella Bianca. (G.B. Guidotti)

The pre-series 2500 C (Coloniale) with camouflage matt paint.

The 2500 C series version. The additional rear tanks are of the type, incorporated into the bodywork so as not to break the profile.

General Gastone Gambara, Chief of Staff of the armed forces Higher Command in North Africa, here in the region of Cyrenaica in 1941.

General Francesco Zingales, commander of XX Corpo d'Armata (20th Army Corps), visits the command of the 6th Grenadier Battalion in Libya in the winter 1941–1942. (ACS)

The trials led to some modifications and more tests, and the definitive 2500 C (Coloniale) entered service in 1941. In North Africa the car was less than completely satisfactory, receiving complaints regarding its suspension and tyres, unlike service in Russia (closed cab variant) where it was judged to be suitable for long trips over unimproved roads, although there was some dissatisfaction with its suspension in Russia as well. Among others, the 2500 C was used by Generals Gastone Gambara and Erwin Rommel in North Africa, by Prince Umberto of Savoy and by Mussolini himself. Production continued until the end of war resulting in a total of 194 machines (150 for the *Regio Esercito* in 1941 and 1942, plus 44 for the occupying German forces in 1944 and 1945), plus the two prototypes.

Technical Description

The civilian Alfa Romeo 2500 6C (six cylinders) was a luxury car by the standards of the day, and the derivative 'colonial' version 2500 C was an equally impressive car.

The Alfa 2500 C was a four-door open touring car (torpedo, according to the period nomenclature) with a retractable canvas roof and had a dividing partition between the driver's compartment and the passenger compartment; the bodywork was made by the coachbuilder Castagna. It was a 4x2 car with rear wheel drive and right-hand driven. Pneumatic tyres were mounted on pressed steel wheels with 14 holes around the perimeter of the rim; two spare wheels and tyres were mounted between the front doors and the front fenders, partially nested in the front fender, on each side. The frame was strengthened, and the wheelbase shortened compared to the civilian model; the independent suspension had helical springs in front and longitudinal torsion bars in the rear. The differential was lockable by the driver. The brakes were hydraulic, drum type on the four wheels. Lubrication and cooling systems were designed to allow the engine to operate in a high-temperature environment, and the carburettor air intake

The Alfa Romeo 2500 C used by Field Marshal Erwin Rommel. Unfortunately the images are of bad quality and some markings have been censored.

Alfa Romeo 2500 C and other vehicles in the Donetsk airfield, in 1942. On the left a S.81 Pipistrello bomber/cargo aircraft is visible.

The 2500 C sedan version on the Eastern Front.

was fitted with an oil-bath filter. The batteries for the electric starter were placed underneath the seats to protect them from heat. The 2500 C had four reserve fuel tanks; two of 25 litres each behind the rear fenders and two of ten litres each, triangular in shape, behind the spare wheels. The engine was a six-cylinder gasoline engine developing 87 HP. The manual transmission had four forward and one reverse gears, with third and fourth synchronised, and a locking differential.

Variants

An enclosed version (sedan) of the Alfa Romeo 2500 C was developed for use in Russia and the Balkans; 50 examples were built, using the chassis available for the torpedo version.

In 1943, Alfa Romeo proposed a 4x4 version of the 2500 C, with a reduction gear, but the proposal was not accepted.

Specifications

- Designation: Autovettura Alfa Romeo 2500 C
- Producer: Alfa Romeo, Milan
- Years produced: 1939–1945
- Number produced: 196
- Length: 4,700mm (15ft 5in)
- Width: 1,570mm (5ft 2in)
- Height: 1,600mm (5ft 3in)
- Unladen weight: 1,725kg (3,803lbs)
- Carrying capacity: Driver and four/six passengers
- Wheelbase: 3,100mm (10ft 2in)
- Front track: 1,445mm (4ft 8in)
- Rear track: 1,447mm (4ft 8in)
- Minimum turning radius: 5,750mm (18ft 10in)
- Minimum clearance: 260mm (10.23in)
- Tyres: Superflex 5.50 x 18
- Engine: Alfa Romeo, six-cylinder, water-cooled, 2,443cc, 54 HP @4,600 rpm
- Fuel: Gasoline
- Transmission: Four speeds forward, one reverse; 3rd and 4th gears synchronised
- Fuel capacity: 120 + 70 reserve in four supplementary tanks, 190 litres total (50 US gallons, 41.8 Imperial gallons)
- Drive layout: 4x2
- Maximum speed: 127km/h (79mph)
- Range (on road): 850km (528 miles) with supplemental fuel tanks

The Alfa Romeo 2500 C used by Field Marshal Erwin Rommel. Unfortunately the images are of bad quality and some markings have been censored.

Alfa Romeo 2500 C and other vehicles in the Donetsk airfield, in 1942. On the left a S.81 Pipistrello bomber/cargo aircraft is visible.

The 2500 C sedan version on the Eastern Front.

was fitted with an oil-bath filter. The batteries for the electric starter were placed underneath the seats to protect them from heat. The 2500 C had four reserve fuel tanks; two of 25 litres each behind the rear fenders and two of ten litres each, triangular in shape, behind the spare wheels. The engine was a six-cylinder gasoline engine developing 87 HP. The manual transmission had four forward and one reverse gears, with third and fourth synchronised, and a locking differential.

Variants

An enclosed version (sedan) of the Alfa Romeo 2500 C was developed for use in Russia and the Balkans; 50 examples were built, using the chassis available for the torpedo version.

In 1943, Alfa Romeo proposed a 4x4 version of the 2500 C, with a reduction gear, but the proposal was not accepted.

Specifications

- Designation: Autovettura Alfa Romeo 2500 C
- Producer: Alfa Romeo, Milan
- Years produced: 1939–1945
- Number produced: 196
- Length: 4,700mm (15ft 5in)
- Width: 1,570mm (5ft 2in)
- Height: 1,600mm (5ft 3in)
- Unladen weight: 1,725kg (3,803lbs)
- Carrying capacity: Driver and four/six passengers
- Wheelbase: 3,100mm (10ft 2in)
- Front track: 1,445mm (4ft 8in)
- Rear track: 1,447mm (4ft 8in)
- Minimum turning radius: 5,750mm (18ft 10in)
- Minimum clearance: 260mm (10.23in)
- Tyres: Superflex 5.50 x 18
- Engine: Alfa Romeo, six-cylinder, water-cooled, 2,443cc, 54 HP @4,600 rpm
- Fuel: Gasoline
- Transmission: Four speeds forward, one reverse; 3rd and 4th gears synchronised
- Fuel capacity: 120 + 70 reserve in four supplementary tanks, 190 litres total (50 US gallons, 41.8 Imperial gallons)
- Drive layout: 4x2
- Maximum speed: 127km/h (79mph)
- Range (on road): 850km (528 miles) with supplemental fuel tanks

Autovettura Bianchi VM6 C

The series Bianchi VM6 C (Coloniale). (CSM/AUSSME)

Benito Mussolini and Marshal Pietro Badoglio at the Piccolo San Bernardo, a mountain pass in the Alps on the Italy-France border, on 28 June 1940. The VM6 C has Superflex Stella Bianca tyres. Note the circular badge indicating that the vehicle belongs to the *Regio Esercito* applied at the bottom of the windshield. (ACS)

A frame from a contemporary newsreel. The camouflage, visible on other military cars of the period, was probably in reddish-brown and dark green spots on a regulation grey-green base colour.

The same car as above. Note the dividing screen between the front and rear seats. (ACS)

Developmental and Service History

The Bianchi VM6 C (Coloniale) was derived in 1938 from the civilian Bianchi S6 sedan; it was built in only 100 copies. The VM6 C was intended for use as a staff car for high-ranking officers; it is known to have been in service in Russia, and probably in the Balkans, as well as in Italy itself. Some are known to have had machine guns mounted in the rear. Production ended in 1940.

Technical Description

The Bianchi VM6 C was a 4x2, rear-wheel drive open touring car. Its four half-doors had no glass windows, but there was a dividing screen between the front and rear seats whose ends could be rotated outwards. The squarish front windshield could be folded fully forward.

Compared to the civilian version, the military version had free-wheeling spares mounted behind the front mudguards. As happened in other cases, chassis and suspension had been reinforced. As befitted a vehicle that served as a command vehicle, the rear compartment had two small folding tables and three small storage compartments for maps and papers. Suspension was semi-elliptical springs and hydraulic shock absorbers; all four wheels had hydraulic brakes, and the hand brake acted on the rear

Autovettura Bianchi VM6 C

The series Bianchi VM6 C (Coloniale). (CSM/AUSSME)

Benito Mussolini and Marshal Pietro Badoglio at the Piccolo San Bernardo, a mountain pass in the Alps on the Italy-France border, on 28 June 1940. The VM6 C has Superflex Stella Bianca tyres. Note the circular badge indicating that the vehicle belongs to the *Regio Esercito* applied at the bottom of the windshield. (ACS)

A frame from a contemporary newsreel. The camouflage, visible on other military cars of the period, was probably in reddish-brown and dark green spots on a regulation grey-green base colour.

The same car as above. Note the dividing screen between the front and rear seats. (ACS)

Developmental and Service History

The Bianchi VM6 C (Coloniale) was derived in 1938 from the civilian Bianchi S6 sedan; it was built in only 100 copies. The VM6 C was intended for use as a staff car for high-ranking officers; it is known to have been in service in Russia, and probably in the Balkans, as well as in Italy itself. Some are known to have had machine guns mounted in the rear. Production ended in 1940.

Technical Description

The Bianchi VM6 C was a 4x2, rear-wheel drive open touring car. Its four half-doors had no glass windows, but there was a dividing screen between the front and rear seats whose ends could be rotated outwards. The squarish front windshield could be folded fully forward.

Compared to the civilian version, the military version had free-wheeling spares mounted behind the front mudguards. As happened in other cases, chassis and suspension had been reinforced. As befitted a vehicle that served as a command vehicle, the rear compartment had two small folding tables and three small storage compartments for maps and papers. Suspension was semi-elliptical springs and hydraulic shock absorbers; all four wheels had hydraulic brakes, and the hand brake acted on the rear

Another VM6 C showing the three-tone camouflage. (ACS)

wheels. Starting was electric. The engine was a six-cylinder gasoline engine developing 52 HP. The manual transmission had four forward and one reverse speeds.

Specifications
- Designation: Autovettura Bianchi VM6 C
- Producer: Edoardo Bianchi, Milan
- Years produced: 1938–1940
- Number produced: 100
- Length: 4,247mm (13ft 11in)
- Width: 1,680mm (5ft 6in)
- Height: 1,680mm (5ft 6in)
- Unladen weight: 1,500kg (3,307lbs)
- Carrying capacity: Four passengers
- Wheelbase: 2,785mm (9ft 2in)
- Front track: 1,420mm (4ft 8in)
- Rear track: 1,420mm (4ft 8in)
- Minimum turning radius: 5,120mm (16ft 8in)
- Minimum clearance: 230mm (9in)
- Tyres: Superflex 5.50 x 18
- Engine: Bianchi S9, six-cylinder, water-cooled, 2,179cc, 54 HP @4,000 rpm
- Fuel: Gasoline
- Transmission: Four speeds forward, one reverse
- Fuel capacity: 72 litres (19 US gallons, 15.8 Imperial gallons)
- Drive layout: 4x2
- Maximum speed: 102km/h (63mph)
- Range (on road): 380km (236 miles)

Autovettura Fiat 518 Coloniale

The Fiat 518 C Ardita series (short wheelbase). The body of this car and the other versions was designed by the coachbuilder Pininfarina.

The Fiat 518 L Ardita series (long wheelbase).

The Fiat 518 Coloniale (short wheelbase) also used by the *Regio Esercito*.

An Italian officer in playful pose on a 518 Coloniale.

Vans of the Reparti Fotocinematografici propaganda department following the troops. North Africa, 1941. (ACS)

Developmental and Service History

The Fiat 518 Coloniale civilian car was based on the Fiat 518 Ardita 2000 four-door convertible whose production began in 1933. The colonial version of the 518 was chosen by the army in anticipation of hostilities with Ethiopia to fill the need for a four-passenger automobile with a relatively powerful engine. Details concerning production numbers and service history are unavailable.

Technical Description

The Fiat 518 Coloniale had an open body with four doors; a canvas roof folded back for stowage behind the rear seats. The 518 was a standard configuration 4x2, with rear wheel drive; it was a left-hand driven vehicle. The 518 Coloniale was built both with a short wheelbase (*C Coloniale*, 2,700mm) and a long wheelbase (*L Coloniale*, 3,000mm) as well as the civilian versions 518 C (or *Corta*, short) 518 L (or *lunga*, long). Pneumatic tyres were mounted on pressed steel wheels; spare tyres mounted on wheels were secured on each side of the bonnet. Suspension consisted of solid axles, longitudinal leaf springs and hydraulic shock absorbers; all four wheels had hydraulic drum brakes. A battery and an electric starter were included. The engine was a four-cylinder gasoline engine developing 45 HP. The manual transmission had four forward and one reverse gears.

Variants

On the chassis of the Fiat 518 second series or the 527 Ardita – i.e. the 518 with an engine upgraded to 2,500cc – an unspecified number of vans were built for service with the *Reparti Fotocinematografici* propaganda department following the troops. The staff belonged to the Istituto Luce, an Italian film company founded in 1924 for the distribution of photographs and newsreels.

Specifications

- Designation: Autovettura Fiat 518 Coloniale
- Producer: Fabbrica Italiana Automobili Turin (Fiat), Turin
- Years produced: 1933–1938
- Number produced: Not Known
- Length: 4,241mm (13ft 3in) 518 C Coloniale; 4,542mm (14ft 11in) 518 L Coloniale:
- Width: 1,670mm (5ft 6in)
- Height: 1,670mm (5ft 6in)
- Unladen weight: 1,185kg (2,612lbs) 518 C Coloniale; 1,259kg (2,775lbs) 518 L Coloniale
- Carrying capacity: Four/six passengers
- Wheelbase: 2,700mm (8ft 10in)
- Front track: 1,180mm (3ft 10in)
- Rear track: 1,180mm (3ft 10in)
- Minimum clearance: 250mm (9.84in)
- Minimum turning radius: 5,800mm (19ft)
- Tyres: Superflex 5.50 x 20
- Engine: Fiat Type 118, four-cylinder, side-valve, water-cooled, 1,944cc, 45 HP @3,600 rpm
- Fuel: Gasoline
- Transmission: Four speeds forward, one reverse
- Fuel capacity: 62 litres (16.4 US gallons, 13.6 Imperial gallons)
- Drive layout: 4x2
- Speed: 82km/h (51 mph)

Autovetturetta Fiat 508 M

Developmental and Service History

In 1931 the Italian military authorities set out a request for a small model to be used as staff car and liaison vehicle. Based on this request, the next year Fiat offered a military version of a new model ready to be launched, the Fiat 508 compact car, colloquially known as the Balilla, in a two-door, two-seater version (aka Spider). Note that the civilian Balilla was built in many models over time: first series, the 508 A with three-speed transmission (later called the 508 A); second series, with a four-speed transmission (from 1934) called the 508 B; and a third and final series (the 508 C, from 1937) with a new body and engine which was later restyled and renamed the 1100 (1939).

The civilian first series Fiat 508 Balilla Torpedo (three-speed aka 508 A).

The civilian first series (three-speed) Fiat 508 A Spider. Note the two doors featured on this sporty model and the spare wheel fixed at the back.

A first-series 508 A Spider with *Regio Esercito* number plate in 1940. Compared to the previous photo, it has stamped disc wheels and the front bumpers are missing.

The 508 M two-door, two-seater, aka Spider Militare, military series version derived from the civilian 508 A Spider. Compared to this latter and the lack of bumpers, the larger and squared rear trunk, the stamped disc wheels with bigger tyres, and the spare wheel placed on the left side, are all clearly visible.

The civilian second series Fiat 508 Torpedo (four-speed, aka 508 B), recognisable by the slanted air intakes of the bonnet. This example mounts standard Pirelli Superflex Cord 4.00 x 17.

The 508 M, as the military version was designated, was acquired by the *Regio Esercito* in mid–1932, and the initial issue was to motorised division headquarters. The 508 M was first built on the basis of the 508 A (first series), then of the 508 B (second series). In addition to the *Spider Militare* (two-seat roadster, 1932), more different body styles were designed: *Torpedo Militare*, (four-seat touring, 1934), *Camioncino* (pick-up truck, 1933), and *Furgoncino* (small van, 1933).

The *Regio Esercito* acquired the Torpedo Militare and the Spider Militare to also assign them to the officers of the *Reali Carabinieri* (military police), from the rank of Major upwards if they did not have other service vehicles at their disposal. The greatest use of these cars was in the colonies of East Africa, where the distances to be covered were large.

Although production ceased in 1937, being replaced by the 508 CM based on the 508 C (third series) Balilla, many 508 M cars saw service in Italian East Africa and later in Spain, during the Civil War, and during the Second World War.

The 508 M and its variants was judged to be exceptionally stable, easy to drive and had good road handling qualities.

A 508 Torpedo Militare second series (four-speed). (D. Zambon)

The 508 M Spider second series, two-door, two-seater version. In both versions of the Spider Militare (three-speed and four-speed) a seat could be found in the trunk for a third passenger. The spare wheels were therefore placed at the body sides.

A camouflaged second series 508 M Spider.

Technical Description

The 508 M Spider Militare was a small two-seat open passenger car with a rear 'rumble seat' for a third passenger. It had a canvas top which could be deployed manually over the passenger compartment. Pneumatic tyres were mounted on pressed steel wheels; there were two spare wheels/tyres mounted on the sides of the bonnet. The windshield was of a one-piece design.

Compared to the civilian Spyder, the 508 M had larger diameter wheels and a modified transmission for greater traction in order to be able to climb steeper hills; offsetting that was the reduction in speed from 80–85km/h to 72–75km/h, and weight was slightly heavier.

The second series differed from the first in the shape of its radiator grille, the slanted rather than vertical bonnet cooling louvres, the wheelbase was 50mm longer and, importantly, a four-speed gearbox (with 3rd and 4th gears synchronised) replacing the three-speed gearbox.

From left to right: a 508 Furgoncino, a 508 Camioncino and a 508 Spider, all civilian versions first series (three-speed).

A 508 M Spider, probably in French territory in the summer of 1941.

This photograph of a 508 M Spider allows the fully raised top, including the sides, and the rear of the bodywork with the luggage trunk which, once opened, served as a folding seat, to be seen.

All four wheels had drum brakes. Suspension was leaf springs front and rear. The engine was a four-cylinder gasoline engine developing 20 HP in the 508 A and 24 HP in the 508 B.

Variants

The first series was followed by a second series (see above) with a slightly different body style and mechanical components.

The Torpedo Militare version launched in 1934 had four doors and could carry four – a driver and three passengers.

The 508 M Camioncino and Furgoncino variants are described in the chapter on Light Trucks, below.

The Coloniale version (i.e. with characteristics suitable for tropical and desert areas, at least on paper) of the different civilian Balillas was also used by the Italian armed forces and designated 508 M Col. or 508 M Coloniale.

Specifications

(Unless specifically noted, all measurements apply to 508 M aka Spider Militare version)
- Designation: Autovetturetta Fiat 508 M

A 508 Spider of the Royal Hungarian Army in 1939. Note the position of the passenger seated on the rear folding seat. (Fortepan 57463)

- Producer: Fabbrica Italiana Automobili Turin (Fiat), Turin
- Years produced: 1932–1939
- Number produced: Not Known
- Length: 3,250mm (10ft 8in)
- Width: 1,380mm (4ft 8in)
- Height: 1,450mm (4ft 9in) with top down; 1,600mm (5ft 3in) with top up
- Unladen weight: 690kg (1,521lbs)
- Carrying capacity: 240kg (529lbs)
- Wheelbase: 2,250mm (7ft 4in) 508 A; 2,300mm (7ft 6in) 508 B
- Front track: 1,180mm (3ft 10in)
- Rear track: 1,200mm (3ft 11in)
- Minimum turning radius: 4,750mm (15ft 7in)
- Minimum clearance: 220mm (9in)
- Tyres: Pirelli Superflex 4.00 x 17 or Michelin Stop 4.00 x 17 or Superconfort 400 x 135 on first series (508 A) and second series (508 B civilian); Superflex 4.40 x 19 on military 508 M; Aerflex 5.00 x 16 on 508 Coloniale
- Engine: Fiat 108 M four-cylinder, side-valve, water-cooled, 995cc, 20 HP @3,400 rpm 508 A; 24 HP @3,800 rpm 508 B
- Fuel: Gasoline
- Transmission: three speeds forward, one reverse 508 A; four speeds forward, one reverse (3rd and 4th gears synchronised) 508 B
- Fuel capacity: 26 litres (6.9 US gallons, 5.7 Imperial gallons)
- Drive layout: 4x2
- Speed: 75km/h (46.6mph)
- Range (on road): 290km (180) 508 A and 508 B; 310km (193 miles) 508 M

Autovettura Fiat 508 C and 508 C Coloniale

The Fiat 508 C (the third model of the civilian Balilla), Berlina or sedan version.

Two examples of civilian 508 C Balilla Berlina cars in service with the Italian forces in Libya in September 1940, and the Ukraine in August 1941. (ACS)

A Balilla Berlina with number plate of the *Regia Aeronautica* in 1941.

Balilla Berlina with number plate of the *Regio Esercito*.

Developmental and Service History

The Nuova Balilla Fiat 508 C (third series of the Balillas), civilian version, began to be produced in 1937 in several configurations – sedan, touring, convertible and cabriolet (by Viotti).

74 ITALIAN SOFT-SKINNED VEHICLES OF THE SECOND WORLD WAR

Two Balillas of different bodies in service with the Regio Esercito. Note the different shape of the number plates.

The Fiat 508 C Coloniale, a four-door convertible intended for use in overseas colonies.

In 1939 the 508 C underwent a restyling, mainly in the grille and in the bonnet; from then on its name became the *Millecento* (one thousand and one hundred) or 1100 in reference to its 1,089 (1,100) cubic centimetre engine. This model was also produced with limited modifications, under licence in France by Simca and in Germany by NSU.

The redesigned version 1100, because of the bulky front, was nicknamed *Musone* (big snout). So, the Nuova Balilla 508 C early model was afterwards christened *Musetto* (small snout).

Initially, the Italian armed forces incorporated both the 508 C Berlina, i.e. the sedan civilian version with four seats and four doors, and its colonial variant, a convertible model slightly modified for the use in the African colonies. The version for military purposes was based on the colonial civilian variant, with a few modifications, and for this reason it was christened 508 C Militare Coloniale (in short, 508 Mil. Col.). It featured two spare wheels at the rear. This car finally represented a transitional model and only a few examples were produced for the *Regio Esercito*. But their positive features, namely the low fuel consumption rate, high stability and good speed, despite traction only on the rear wheels, prompted the military authorities to ask Fiat for a new torpedo all-terrain version.

So, in mid–1938 the definitive model of the military car was produced, which was renamed the 508 C Torpedo Militare or, more briefly, the 508 CM (i.e. third series, military). This is described below.

Technical Description

The civilian Nuova Balilla, initialled 508 C, featured a completely different bodywork from those of the two previous series (508 A and 508 B, as they were called from that moment on). Even the engine, the Fiat 108C, was redesigned and was more powerful. The gearbox had four gears, plus reverse, and the car could reach 110km/h. Another difference compared to the previous series

Two views of a Fiat 508 C Militare Coloniale. It was almost identical to the civilian coloniale, except for the two rear spare wheels.

The Fiat 1100 Musone, i.e. the restyled 508 C civilian sedan.

lay in the suspension: the front suspension had independent wheels equipped with coil springs assisted by hydraulic shock absorbers; the rear ones remained of the leaf spring type. The 1100, apart from the bodywork, was basically identical.

Variants
A long wheelbase version, called 508 L and capable of carrying six passengers, was also manufactured.

Specifications
- Designation: Autovetturetta Fiat 508 C
- Producer: Fabbrica Italiana Automobili Turin (Fiat), Turin
- Years produced: 1938–1953
- Number produced: more than 57,000 examples

- Length: 3,615mm (11ft 10in)
- Width: 1,480mm (4ft 10in)
- Height: 1,480mm (4ft 10in); 1630mm (5ft 4in) with top closed
- Total weight: 1,160kg (2,557lbs)
- Carrying capacity: 4 passengers – 300kg (661lbs)
- Wheelbase: 2,420mm (7ft 11in)
- Front track: 1,231mm (4ft ½in)
- Rear track: 1,226mm (4ft)
- Minimum turning radius: 4,500mm (14ft 9in)
- Minimum clearance: 170mm (6.7in)
- Tyres: Superflex 5.00 x 15 Berlina; 5.00 x 17 Coloniale
- Engine: Fiat 108 C, four-cylinder, valve-in-head, water-cooled, 1,089cc, 32 HP @4,400 rpm
- Fuel: Gasoline
- Transmission: Four speeds forward, one reverse
- Fuel capacity: 32 litres (8.45 US gallons, 7 Imperial gallons)
- Drive layout: 4x2
- Speed: 110km/h (68mph)
- Range (on road): 370km (230 miles)

Autovettura Fiat 508 CM (1100 Mimetica)

Note

To avoid confusion with the models described above, in this entry the denominations 508 C Torpedo Militare and 508 C Torpedo Coloniale have been used, instead of 508 Militare and 508 C Militare Coloniale, although the latter two are also present in the documentation issued by Fiat.

The prototype of the Fiat 508 CM. (Fiat)

Developmental and Service History

As previously mentioned, towards the end of the thirties the Italian military authorities asked Fiat to design a new off-road car in a torpedo configuration more suitable for use in war zones than the 508 M, which had nonetheless impressed.

The Turin company started from the chassis of the new Balilla, i.e. the third series or Fiat 508 C (see above). In mid-1938, the civilian car body was completely redesigned to meet military specifications and requirements thus taking on a much more 'martial appearance'. This became the definitive version of the military car, which was renamed the 508 C Torpedo Militare or, more briefly, the 508 CM (i.e. third series, military). It had a sturdier chassis, reinforced suspension, increased ground clearance and modified transmission with a top speed limited to 95km/h. The wheels were also larger. But the most distinctive element was the squared-off body. The early definitive 508 CM had a camouflage paint scheme, and – considering the change of name of the civilian version – was thus often referred to as Fiat 1100 Mimetica (camouflaged).

COLOUR SECTION

Fiat 508 CM (aka Fiat 1100 Mimetica) staff car, Italy, 1939. (Artwork by and © David Bocquelet)

Autocarretta 35 light truck, mountain chain of the Alps, Italy, 1935. (Artwork by and © David Bocquelet)

Fiat 618 light truck, East Africa, 1936. (Artwork by and © David Bocquelet)

SPA 38R light truck, Eastern Front, 1941. (Artwork by and © David Bocquelet)

SPA CL 39 light truck, Eastern Front, 1941. (Artwork by and © David Bocquelet)

SPA AS 37 (second series) light truck, North Africa, 1942. (Artwork by and © David Bocquelet)

Benito Mussolini and Marshal Pietro Badoglio on board a Bianchi VM6 C staff car in a newsreel frame from 28 June 1940.

Photograph taken by a German soldier during the transfer trip and stay in North Africa. On the left, an Opel Olympia 38 that appears camouflaged in the *Tropen* colours employed by the Wehrmacht since March 1942. On the right, a Fiat 508 C Camioncino belonging to the *Regio Esercito* whose grey-green colour is only partially covered by the *kaki sahariano* (sand-yellow) applied by spray gun.

A 508 L Camioncino, painted grey-green.

An advertising poster for the Breda company (1931).

The cover of a 1934 motorcycling magazine with Sertum company propaganda.

The updated version of the Viberti brand.

The original brand of coachbuilder Viberti.

The Fiat 508 CM, aka 508 C Torpedo Militare, aka Fiat 1100 Mimetica. (Fiat)

A 508 CM on the assembly line in 1940. (LIFE)

The 508 CM / 1100 Mimetica was produced from 1938 to 1945 and saw service on all fronts in which the Italians were engaged. It was generally issued to the motor pools of headquarters elements and was extensively used by officers. It was the most widely used liaison and reconnaissance vehicle in the *Regio Esercito* and the *Regia Aeronautica*. About 6,000 examples of the 508 CM were built until 1943, and a further 1,354 examples were delivered to the German forces after the Armistice. The surviving vehicles served in Italy until 1960.

Technical Description

The 508 CM, aka 1100 Mimetica, was a four-door, four-seater, 4x2, right-hand, rear-wheel drive car. It was normally seen used as an open car, but it did have a retractable canvas roof and side curtains, with clear window inserts made of celluloid. The windshield was of a one-piece type with a single electric windshield wiper for the driver. The wheels were pressed steel discs with

Factory photo of the Fiat 508 CM or 1100 Mimetica. (Fiat)

An artillery officer aboard a Fiat 508 CM aka 1100 Mimetica.

Mussolini reviews the troops aboard a Fiat 1100 Mimetica.

Fiat 1100 Mimetica. (D. Zambon)

Bir El Gubi, Libya, autumn 1941. A Fiat 508 CMC, aka Fiat 1100 Torpedo Coloniale, together with motorcycles, trucks and AB 41 armoured cars. The 'balloon' tyres and the yellow sand colour confirm that this is a colonial variant, (ACS)

A small 4x2 car like the 508 CM was unable to cope with the mud of the Eastern Front. (ACS)

small hub caps; a spare wheel and tyre were bolted onto the rear of the body. The chassis consisted of two longerons of pressed sheet metal; it was narrower in the front and was made rigid by a double cross-member in the shape of an 'X'. The front bumper was attached to the longerons. The independent front suspension consisted of helical springs with hydraulic shock absorbers; rear suspension consisted of a solid rear axle, leaf springs, hydraulic shocks and a stabiliser bar. Brakes were four-wheel hydraulic drums and a hand brake acting on the transmission. The electrical system was a 12 volt system with battery and electric starter. The engine was a four-cylinder gasoline engine developing 30 HP. The manual transmission had four forward and one reverse gears.

Variants

The 508 CM conceived for the North African theatre, named 1100 Torpedo Coloniale or 508 CMC, manufactured until 1943, was fitted with special filters for the desert environment, larger fuel tank (70 litres instead of 40 litres), enhanced battery, modified transmission, different wheel track and ground clearance. The standard, low pressure Superflex Cord 5.00 x 18 pneumatic tyres were replaced with very low pressure Aerflex 6.00 x 16 balloon tyres, smaller but wider and thus more suitable for soft and sandy terrain. Except for these elements and the colour, which was the *kaki sahariano* (sand-yellow) specific to that theatre of war, it was almost identical to the standard 1100 Mimetica.

Siliana, Tunisia, April 1943. A posed photo of French Goumiers of the 2eme Tabor of the 1st GTM (Group of Moroccan Tabors) of the 1st DMM (Moroccan Marching Division) aboard a captured FIAT 508 CM. (ECPAD)

The Fiat 508 CM Berlina FO (*Fronte Orientale*, i.e. Eastern Front) was a completely enclosed and heated sedan, resembling the civilian model. 50 examples were assigned to the Italian expeditionary corps in Russia in 1941.

The 508 CM frame was used as the basis for the Fiat 1100 Camioncino pick-up truck (see below).

Specifications

- Designation: Fiat 508 Torpedo Militare or Fiat 508 CM
- Producer: Fabbrica Italiana Automobili Turin (Fiat), Turin
- Years produced: 1938–1945
- Number produced: more than 7,000 examples
- Length: 3,615mm (11ft 10in)

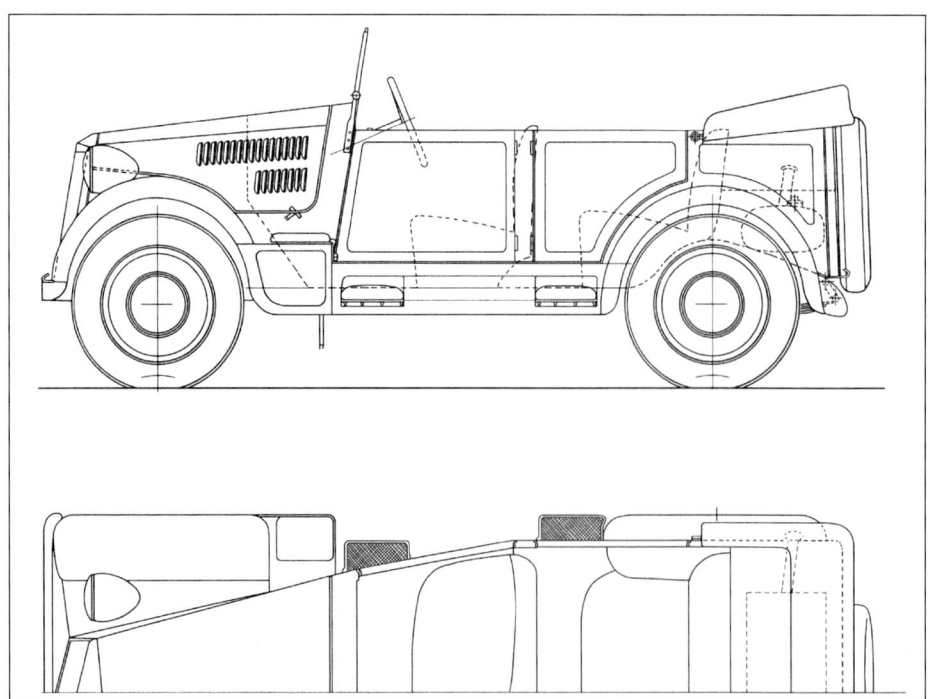
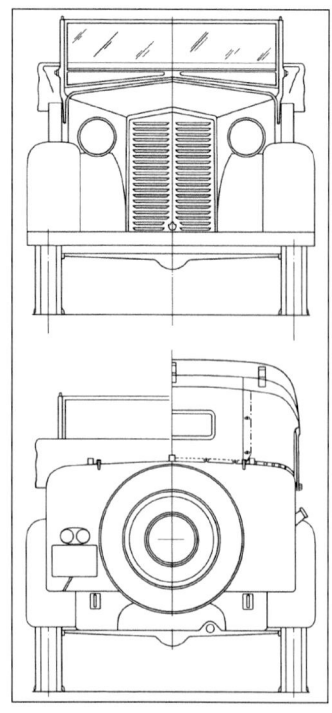

Fiat 508 CM. (Drawings by GMT)

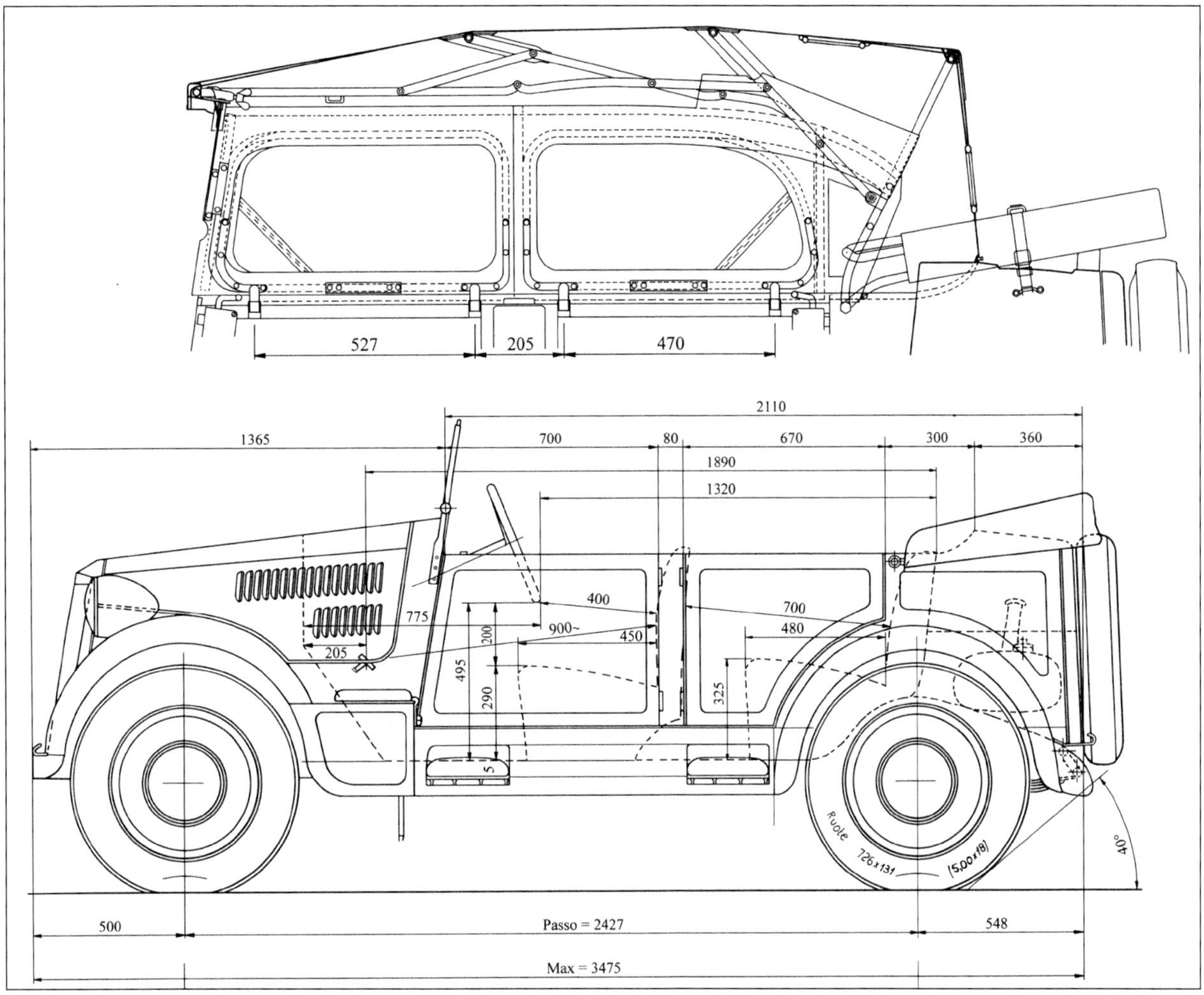

Original Fiat drawings with dimensions of the car and the retractable canvas roof curtain (left side).

Fig. 1. — Autotelaio Mod. 508 C Militare.

Transparent view taken from the user manual.

82 ITALIAN SOFT-SKINNED VEHICLES OF THE SECOND WORLD WAR

We do not know if this Fiat 508 CM Coloniale with special equipment, conceived for special units similar the British Long Range Desert Group (LRDG), was ever employed or was just a proposal. It carries pioneer tools, additional canisters, a searchlight on the left side but lacks a windshield. Additionally, it is armed with a 6.5mm Breda 30 light machine gun. Inside, the barrels of some other weapons are visible, including that of a MAB 38 A submachine gun. (C. Pergher)

- Width: 1,480mm (4ft 10in)
- Height: 1,480mm (4ft 10in); 1630mm (5ft 4in) with top closed
- Unladen weight: 890kg (1,962lbs)
- Carrying capacity: 300kg (661lbs)
- Wheelbase: 2,427mm (7ft 11in)
- Front track: 1,242mm (4ft) Torpedo Militare; 1,256mm (4ft 2in) Torpedo Coloniale
- Rear track: 1,239mm (4ft) Torpedo Militare; 1,253mm (4ft 1in) Torpedo Coloniale
- Minimum turning radius: 5,000mm (16ft 5in)
- Minimum clearance: 237mm (9¼in) Torpedo Militare; 230mm (9in) Torpedo Coloniale
- Tyres: Superflex 5.00 x 18 Torpedo Militare; Aerflex 6.00 x 16 Torpedo Coloniale
- Engine: Fiat 108 C, four-cylinder, valve-in-head, water-cooled, 1,089cc, 30 HP @4,400 rpm
- Fuel: Gasoline
- Transmission: Four speeds forward, one reverse
- Fuel capacity: 40 litres (10.57 US gallons, 8.8 Imperial gallons) Torpedo Militare; 70 litres (18.5 US gallons, 15.4 Imperial gallons) Torpedo Coloniale
- Drive layout: 4x2
- Speed: 95km/h (59mph)
- Range (on road): 320km (201 miles) Torpedo Militare; 570km (354 miles) Torpedo Coloniale

Autovetturetta Fiat 500 Topolino

Developmental and Service History

Written and photographic documentation tell us that the small Fiat 500 car was never officially adopted by the Italian armed forces. However, some examples may have been purchased by the Italian Army, Air Force or Navy for assignment to officers or for liaison duties in metropolitan areas. Although, probably, other cars – such as the Fiat 508 in all its variants – were preferred.

The 500 project started in 1934; the Turin-based manufacturer felt confident from its experiences with the 509 and 508 models and decided to create 'a small two-seater capable of making motoring more popular in Italy', as the owner Giovanni Agnelli declared. The story that it was Mussolini who personally requested such a small and cheap car is unconfirmed, but it nevertheless falls within the climate of that period.

The Fiat 500, launched in 1936, was later nicknamed *Topolino*, because of its minute and tapered shape – the Italian translation of 'Mickey Mouse' (*Topolino*) actually had nothing to do with the car, although it later helped to spread the name more widely.

As happened with almost all the Italian civilian and military vehicles, a great number of examples were requisitioned by the Germans after the Armistice. This model is also called Fiat 500 A to differentiate it from the 500 B and 500 C models produced after the end of the war.

The Fiat 500 was also produced in France by Simca (with the name Simca 5), in Poland by Polski-Fiat and in Germany by NSU, all with small external and internal differences in comparison to the original Italian model.

A civilian Fiat 500 Topolino in front of the Ente Italiano per le Audizioni Radiofoniche, the Italian National Radio Broadcasting Station in Asiago Street, during the operations in Rome following the Armistice of 8 September 1943.

A Fiat 1100 pick-up and a Fiat 500 Topolino Trasformabile (with folding top) employed by 3. Batterie, I. Abteilung, Artillerie-Regiment Hermann Göring, LW Panzer-Division Hermann Göring.

A 500 Topolino used by Stug-Kompanie of the Pz.Jg.Abt. 46, Reichsgrenadier-Division Hoch und Deutschmeister in late 1943. The vehicle received new markings, number plate and camouflage from the new owners.

A Fiat 500 Topolino from the German 26. Panzer-Division in Formignana, near Ferrara, in January 1945. Note the 'FIAT' logo on the wheel rim cover. It is an example produced after 1938, with reinforced rear suspension, turn indicators on the windscreen struts and the interior rear mirror at the top. Some details, such as the external mirror and the two-piece bumpers – not included in the standard models in Italy, see above – suggest a special version or one produced for the foreign market.

A huge German Büssing-NAG Typ 650 together with a Fiat 500 A Topolino car. Both vehicles belonged to a headquarters and shown a complete multi-coloured camouflage. The car had an improvised, but official, German number plate.

Milan, a Repubblica Sociale Italiana column – which includes a Bianchi Miles truck and an M40 tank – at the halt among Italian and German soldiers. The Fiat 500 Topolino was a common sight during the Italian Campaign. (S. Di Giusto)

The Fiat 500 short leaf spring chassis furgoncino (small van). (Fiat)

Technical Description

The first model produced, the Fiat 500, could reach a speed of 85km/h and carry two adults and a 50kg bag or two children. It weighed 535kg and, thanks to a 569cc engine, consumed 6 litres of gasoline per 100km. The chassis consisted of two longerons of pressed steel plate, with lightening holes, connected by two cross-members. The front wheels had independent suspension with upper transverse leaf spring and hydraulic shock absorbers. The suspension of the rear wheels were half-spring in the first series (*balestra corta*, short spring), and full leaf springs in the second series (*balestra lunga*, long spring) with hydraulic shock absorbers.

Variants

The Fiat 500 Trasformabile (convertible) had a folding top.

A van with a capacity of 300kg (Furgoncino) was built on the chassis of the Fiat 500. The version on short leaf spring chassis had only one inclined rear door, the version on the long leaf spring chassis, reinforced compared to the previous one, had two vertical rear doors.

A Fiat 500 long leaf spring furgoncino of the Fallschirm-Sanitäts-Abteilung 4 from 4. Fallschirmjäger-Division, a unit formed in Italy in November 1943. The white rampant horse within a black circle is visible on many Italian vehicles used by the German Medical Corps, which were repainted completely white.

The Fiat 500 A Topolino. (Drawings by A. M. Feller – GMT)

Specifications
- Designation: Autovetturetta Fiat mod. 500
- Producer: Fabbrica Italiana Automobili Turin (Fiat), Turin
- Years produced: 1936–1948
- Number produced: 122,213
- Length: 3,215mm (10ft 6in)
- Width: 1,275mm (4ft 2in)
- Height: 1,377mm (4ft 6in)
- Unladen weight: 535kg (1,180lbs)
- Wheelbase: 2,000mm (6ft 7in)

- Front track: 1,114mm (3ft 8in)
- Rear track: 1,083mm (3ft 6in)
- Tyres: Superflex 4.00 x 15
- Engine: Fiat 500 four-cylinder, side-valve, water-cooled, 569cc, 13 HP @4,000 rpm
- Fuel: Gasoline
- Transmission: four speeds forward, one reverse (3rd and 4th gears synchronised)
- Fuel capacity: 21 litres (5.5 US gallons, 4.6 Imperial gallons)
- Drive layout: 4x2
- Speed: 85km/h (53mph)

Autovettura Fiat 2800 CMC

Benito Mussolini greets King Vittorio Emanuele III aboard his Fiat 2800 Torpedo Reale. (ACS)

The prototype of the Fiat 2800 Corta Militare Coloniale. (Fiat)

The series Fiat 2800 CMC during tests in Libya in 1939. Compared with the prototype, note the spare wheels positioned at the sides, and the differences in the body and retractable curtain. (Fiat)

Developmental and Service History

The Fiat 2800 CMC (the CMC stood for *Corta Militare Coloniale*, or short military colonial) was a military version of the somewhat elegant 2800 four-door touring car produced from 1938 to 1944. The civilian Berlina and Torpedo versions of the car were widely used as official vehicles, and among other notables, were used to carry King Vittorio Emanuele III, Prince Umberto of Savoia, Mussolini and the Pope. Some 624 2800s chassis were built, 210 of which were for Italian military use and intended for high-ranking officers. A few survived the war and served in the post-war Italian Army until replaced by more modern vehicles.

Technical Description

The Fiat 2800 CMC was smaller and incorporated a number of improvements and features not found in the civilian version of the 2800. The 2800 CMC was a four-door, rear-wheel drive touring car (with a body similar to the 508 CM body) with a retractable canvas top. The all-metal body had prominent curved fenders; an early version carried two spare tyres on the rear of the body, while the definitive version moved the two spares to the sides of the bonnet, directly behind and partially sunk into the front fenders. Mechanically, the 2800 CMC had a reinforced shorter frame (3.0 metres instead of 3.2 metres), heightened ground clearance, was fitted with an air filter, oil filters, an auxiliary electric fuel pump, two 6 volt batteries, a modified transmission, bigger tyres, modified hand brake and a transversally mounted muffler; it also had a 130 litre fuel tank. The traction was rear, with four-speed manual gearbox and hydraulic brakes. The independent front suspension consisted of helical springs and hydraulic shock absorbers; rear suspension was a solid axle with longitudinal leaf springs and hydraulic shock absorbers. The engine was a six-cylinder gasoline engine developing 85–90 HP. The manual transmission had four forward and one reverse gears.

Another image of the Fiat 2800 CMC during tests. It carries an unusual civilian licence plate of the Government of Libya. (Fiat)

The 2800 CMC with Prince Umberto on board passes among the troops of the 101ª *Divisione Trieste* on the Alps, in June 1940, during the attack on France.

Variants

At the beginning of 1943 the prospect of a variant of the 2800 CMC with four-wheel drive was considered, but the project was never completed.

Specifications

- Designation: Autovettura Fiat 2800 CMC
- Producer: Fabbrica Italiana Automobili Turin (Fiat), Turin
- Years produced: 1939–1944
- Number produced: 624 chassis (about one-third for the CMC)
- Length: 4,795mm (15ft 9in)
- Width: 1,884mm (6ft 2in)
- Height: 1,377mm (4ft 6in); 1,768mm (5ft 10in) with top closed
- Unladen weight: 1,970kg (4,343lbs)
- Carrying capacity: four/six (driver and three passengers, plus two additional jump seats for an additional two passengers)
- Wheelbase: 3,000mm (9ft 10in)
- Front track: 1,452mm (4ft 9in)
- Rear track: 1,460mm (4ft 9in)
- Minimum turning radius: 6,100mm (20ft)
- Minimum clearance: 245mm (10in)
- Tyres: Superflex 4.00 x 18 or 6.50 x 18
- Engine: Fiat 2852 MC six-cylinder, water-cooled, 2,852cc, 90 HP @4,400 rpm
- Fuel: Gasoline
- Transmission: Four speeds forward, one reverse, manual transmission
- Fuel capacity: 130 litres (34 US gallons, 28.6 Imperial gallons)
- Drive layout: 4x2
- Speed: 115km/h (71mph)
- Range (on road): 400km (249 miles)

Autovettura Lancia Aprilia Coloniale

Developmental and Service History

On the eve of the war, the *Regio Esercito* was desperately short of motor vehicles of all types and classes. Lancia offered a militarised version of its Lancia Aprilia sedan car in 1941 and conceived it as being for officers' use. The prototype underwent several modifications, leading to a somewhat simplified definitive version. 251 examples were built between 1941 and 1943.

The Aprilia Coloniale was mainly used by general officers. In service, the car suffered frequent problems with its suspension, steering gear, clutch and radiator. The Aprilia Coloniale saw service on all fronts, and following the 8 September 1943 Armistice, was used to some extent by German forces in Italy.

Technical Description

The Lancia Aprilia Coloniale was a torpedo (touring) car based on the civilian Lancia Aprilia Trasformabile second series, on a modified Lancia 639 S frame, shortened and reinforced in comparison to the civilian model. The body was designed and built

A first series Lancia Aprilia sedan with Fergat 'Littoria' spoked wheels and Michelin tyres.

A Lancia Aprilia Convertible with a Polish licence plate in 1952. (Maciek Peda)

The prototype of the Aprilia Coloniale. (Lancia)

by Viotti, a well-respected coach builder based in Turin. The body itself was a four-door touring type with a retractable canvas top; the window openings were fitted with side curtains and clear window inserts. A feature of the Aprilia Coloniale was a rifle rack behind the driver's seat which could hold four mod. 91 carbines.

The Aprilia Coloniale was a 4x2 vehicle with rear-wheel drive and driven right-hand. Pneumatic tyres were mounted on pressed 'Littoria' type steel disc wheels; a spare wheel and tyre were carried on the rear of the body. The prototype had removable auxiliary fuel tanks, but the production version had 20-litre fuel canisters carried behind the front and rear fenders on each side.

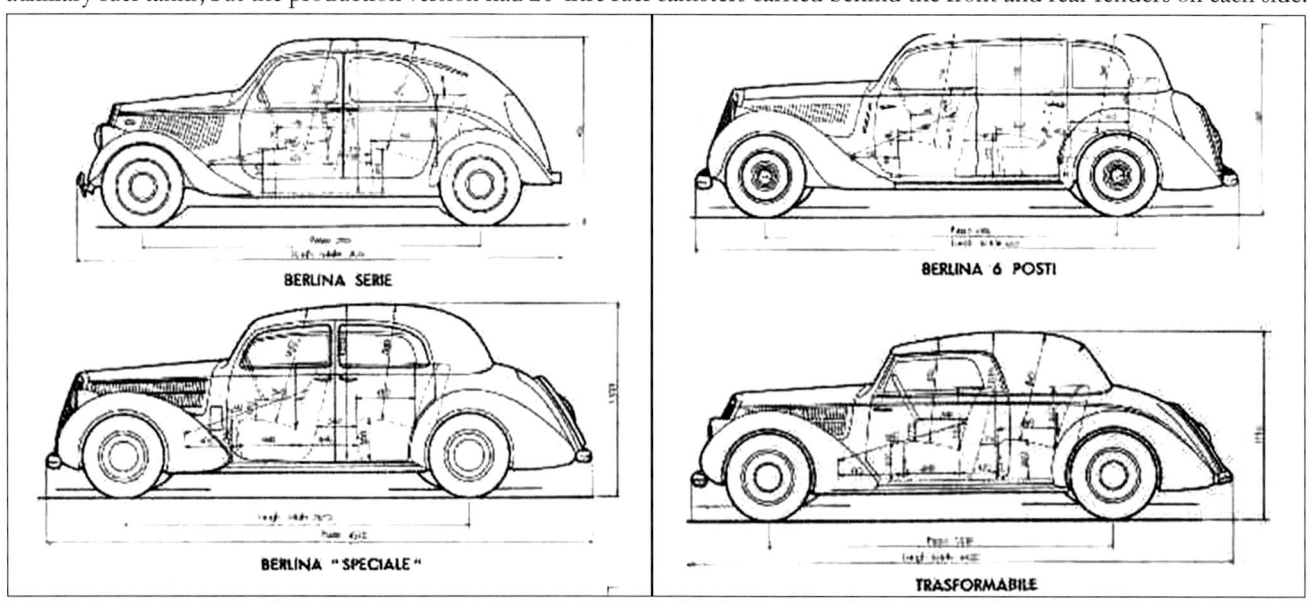

Image taken from a Lancia catalogue which schematically illustrates the four versions of the first series.

Factory pictures of the Lancia Aprilia Coloniale. Note the four 20-litre jerry cans located on the sides, front and rear. The bodywork was very similar to that of the Fiat 508 CM, which sometimes makes it difficult to distinguish the two vehicles in photographs. (Lancia)

The independent front suspension consisted of helical springs and hydraulic shock absorbers; rear suspension was a solid axle with longitudinal leaf springs and hydraulic shock absorbers. The electrical system was a 6-volt system with battery. The engine was a four-cylinder gasoline engine developing 51 HP. The manual transmission had four forward and one reverse gears; the clutch was a single plate dry disc.

The internal rack for rifles and the reclining rear seat. (Lancia)

A Lancia Aprilia Coloniale leading a vehicle convoy crossing a bridge built by Italian engineers in the Ukraine. (ACS)

General Alessandro Gloria, commander of the Divisione Bologna. (G. Forbicini)

Specifications
- Designation: Autovettura Lancia Aprilia Coloniale
- Producer: Lancia & C. Fabbrica Automobili, Turin
- Years produced: 1941–1943
- Number produced: 251
- Length: 4,326mm (14ft 2in)
- Width: 1,550mm (5ft)
- Height: 1,650mm (5ft 5in)
- Unladen weight: 1,112kg (2,452lbs)
- Carrying capacity: Four passengers, including the driver
- Wheelbase: 2,650mm (8ft 8in)
- Front track: 1,233mm (4ft)
- Rear track: 1,292mm (4ft 3in)
- Minimum turning radius: 5,100mm (16ft 9in)
- Minimum clearance: 226mm (9in)
- Tyres: Superflex 13 x 45
- Engine: Lancia Tipo 99 four-cylinder V, water-cooled, 1,485cc, 46 HP @4,000 rpm
- Fuel: Gasoline
- Transmission: Four speeds forward, one reverse
- Fuel capacity: 75 litres, plus four 20-litres fuel canisters (80 litres), total 155 litres (41 US gallons, 34 Imperial gallons)
- Drive layout: 4x2
- Maximum speed: 110km/h (68.3 mph)
- Range (on road): 620km (385 miles) with supplemental fuel canisters

Autovettura Lancia Artena Militare

The civilian Lancia Artena fourth series Ministeriale (#341 chassis) by Pininfarina.

The Lancia Artena Militare 6 luci 4 porte by Viotti.

The Lancia Artena Militare Trasformabile, aka 4 luci 4 porte.

Developmental and Service History

The civilian Lancia Artena was manufactured from 1931 to 1936 in three different production series. The fabrication of a military version, based on the Lancia 341 frame (fourth series), was resumed in 1940 for the *Regio Esercito* to make up for the shortage of military vehicles. Between 1940 and 1942 a total of 507 were built. The car was assigned to high-ranking officers at army group, army and corps level headquarters. Mussolini travelled in an Artena during a visit to the Greek-Albanian Front in 1941.

The Lancia Artena Militare Torpedo Trasformabile with grey-green or camouflage colouring. Note the partially foldable hood.

Technical Description

The Lancia Artena was built on a Lancia 341 frame (fourth series), and included three six-seat body styles:

1) the Berlina, which was a four-door six-window saloon, referred to as the Ministeriale (civilian variant) or the 6 luci 4 porte (six-windows four-door, military variant);

2) the Trasformabile 4 luci 4 porte, a four-window four-door touring car with a folding roof;

3) the Torpedo Trasformabile which was a four-door touring with a more squarish and simplified body, intended for the field use.

Viotti of Turin, which had close ties to Lancia, was commissioned to build the bodies of the military version, while Pininfarina took charge of designing the *Ministeriale*.

The Lancia Artena Militare was a 4x2 vehicle with rear-wheel drive and right-hand drive. Pneumatic tyres were mounted on 'Littoria' style disc wheels made by Fergat, a specialised company based in Turin. A spare wheel and tyre were mounted directly behind the front fenders on each side. The front suspension was independent with hydraulic shock absorbers; rear suspension was a solid axle with semi-elliptical leaf springs and mechanical shock absorbers. All wheels had hydraulic drum brakes. The engine was a water-cooled four-cylinder gasoline engine developing 51 HP. The manual transmission had four forward and one reverse gears; the clutch was a single plate dry clutch. The electrical system used a 12-volt battery.

Variants

As described above. Additionally, 191 were reportedly built with an ambulance body made by Bertone on a different frame – the model 441.

Specifications

- Designation: Autovettura Lancia Artena 4a Serie – Tipo Militare
- Producer: Lancia & C. Fabbrica Automobili, Turin
- Years produced: 1940–1942
- Number produced: 316 (plus 191 with ambulance frame)
- Length: 4,960mm (16ft 3in)
- Width: 1,730mm (5ft 8in)
- Height: 1,750mm (5ft 9in)
- Unladen weight: 956kg (2,107lbs)
- Carrying capacity: six passengers (two of whom were on front facing folding seats in the Trasformabile and Torpedo Trasformabile versions)
- Wheelbase: 3,180mm (10ft 5in)
- Front track: 1,400mm (4ft 7in)
- Rear track: 1,420mm (4ft 8in)
- Minimum turning radius: 5,170mm (17ft 2½in)
- Minimum clearance: 200mm (8in)
- Tyres: Superflex 15 x 45 or 16 x 45
- Engine: Lancia Tipo 84, four-cylinder V, water-cooled, 1,924cc, 51 HP @3,800 rpm
- Fuel: Gasoline
- Transmission: Four speeds forward, one reverse, manual transmission
- Fuel capacity: 70 litres (18.5 US gallons, 15.4 Imperial gallons)
- Drive layout: 4x2
- Maximum speed: 105km/h (65 mph)
- Range (on road): 380km (236 miles)

The Italian armed forces, in particular the *Regia Aeronautica* and, to a lesser extent, the *Regio Esercito* and the *Regia Marina*, employed a number of civilian executive cars such as the Fiat 1500, a model produced from 1935.

Fiat 1500 cars from *Regia Aeronautica* and *Regia Marina*. The three standing officers are Spanish pilots of the Ejército del Aire photographed in Caserta in 1939. The Royal Navy car, with 'RM 2467' number plate and special Pininfarina bodywork, was photographed in La Spezia before the war.

4

Light trucks

Autocarretta OM 32, 35 and 36

The Autocarretta 32, right-hand side view. (CTM)

The Autocarretta 35, left-hand side view. (CTM)

Developmental and Service History

The *autocarretta* (motorised cart) was a uniquely Italian light vehicle, designed expressly for employment in Italy's mountainous terrain, but later used in other roles as well. In 1927 the Italian military authorities set out a request for a vehicle that would be able to operate on bad roads and rough terrain. In fact, one of the official descriptions would later define it as a 'mountain reconnaissance car'.

Prototypes presented by Pavesi and Alfa Romeo were deemed unsuitable, followed by projects from Fiat, Lancia, Ansaldo and Ceirano. The Ansaldo entry was accepted, and a prototype was tested at the end of 1929. However, the *Società Anonima*

Rear view of the Autocarretta 32 (left) and 35 (right) compared. The latter shows a lower height achieved from the wider wheel track and modified suspension. (CTM)

The Autocarretta 35, front view. The central groove of the semi-pneumatic tyres favoured the fastening of the devices to improve grip on the ground. (CTM)

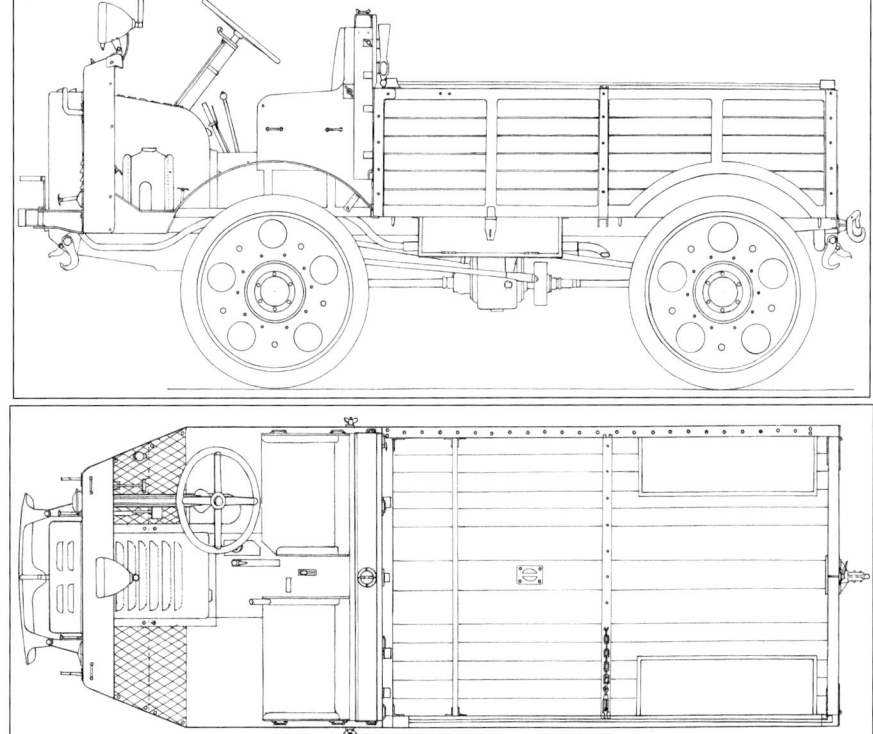

The Autocarretta 35. (Drawings by N. Pignato via GMT)

Automobili Ansaldo, based in Turin (not to be confused with the Genoese industrial group of the same name), was in dire straits and was taken over by the Officine Meccaniche (OM) of Brescia, which also assumed the manufacturing contract for the small vehicle. In the second half of 1931, OM furnished three test vehicles, slightly different from the Ansaldo prototype, which were subjected to severe tests in different conditions, obtaining homologation as the Autocarretta 32 at the beginning of the following year. A batch of pre-production models of these vehicles was issued to units for field testing; some of them took part in the large-scale manoeuvres of August 1932 with great success.

In mid-1932, a second batch of these vehicles in an updated design was ordered, with deliveries completed in 1935. From the applied modifications, dictated by previous experiences, a new model emerged, designed the Autocarretta 35 (or second series). The improvements made the vehicle more stable, thanks to better suspensions and a wider wheel track, and with better

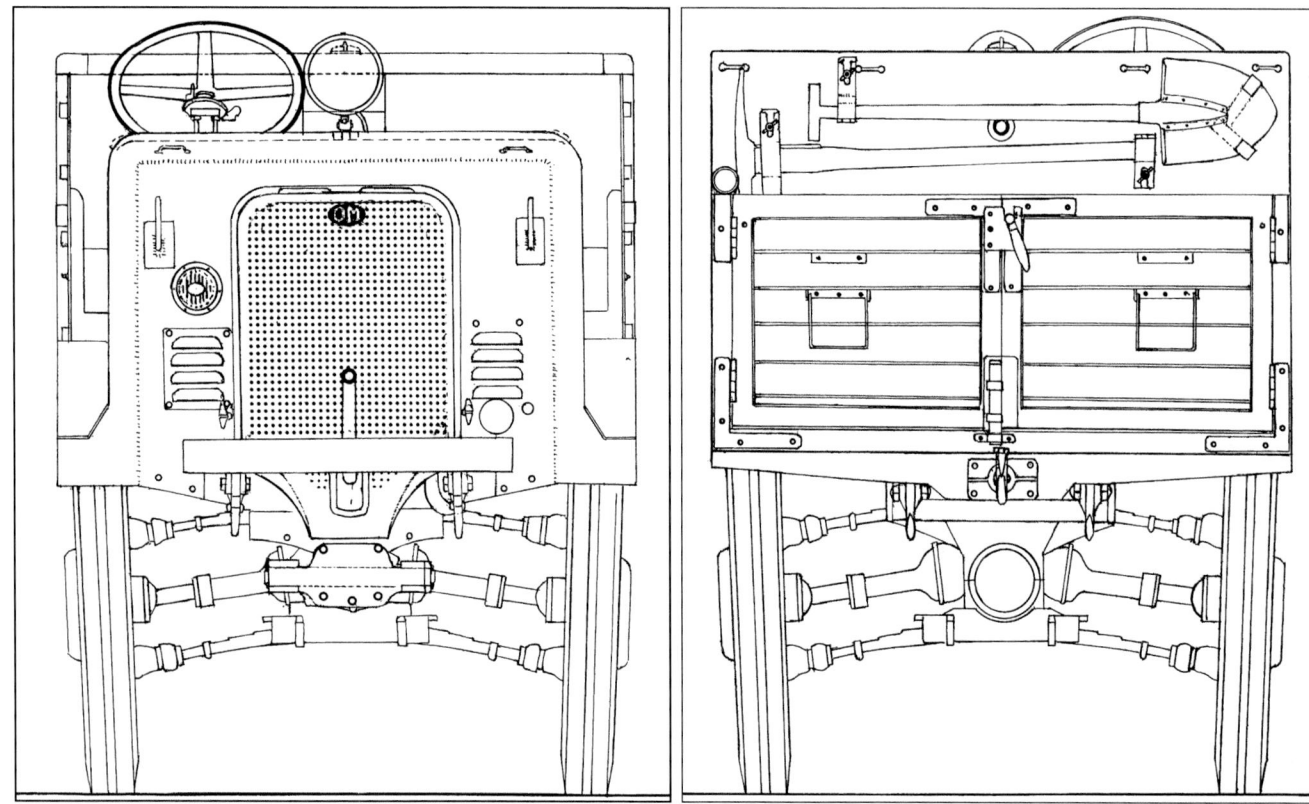

The Autocarretta 35. (Drawings by N. Pignato via GMT)

The *autocarrette* had their baptism of fire in Spain within the *Italian Corpo Truppe Volontarie*, or Voluntary Troops Corps. The camouflage usually included the canvas top, not deployed in the open position here. The local number plate, not always applied, was probably with blue background with yellow letters and numbers. The code 'BS' stands for Base Siviglia (Base Seville). (AUSSME)

road holding, due to the adoption of more efficient steering and transmission components. The model 35, together with the 32 retrofitted with the same modifications, would prove successful in operations in Italian East Africa and Spain in the late 1930s.

Based on field tests with vehicles assigned to a motorised division, a third series was developed, the Autocarretta 36, referred to as the 36 P (personale) – unofficially 36 DM (Divisione Motorizzata) – or 36 Mt (Materiale), depending on the configuration as a troop carrier (P) or a cargo carrier (Mt).

The final series was the Autocarretta 37, adopted shortly before the war, which had the same characteristics as the model 36 Mt but which reverted to semi-pneumatic tyres and lacked a windshield, as well as having minor modifications to the engine and other components.

The Autocarretta 35 and 36 were issued on varying scales to motorised divisions and to the Italian alpine divisions. They served in Italian East Africa (1,366 vehicles were operational as of 30 April 1936) and in Spain during the Civil War (many of which returned from the *Africa Orientale Italiana*); indeed, in 1937, of a total requirement for 9,640 of these vehicles, only 1,800 were available.

In October 1939, on the eve of Italy's entry into the Second World War, 2,751 autocarrette of various models were available on the national territory (excluding the colonies) with another 2,000 on order. They saw service in the Balkans, on the Eastern

The Autocarretta 36 P personnel carrier with Superflex Artiglio pneumatic tyres. Note the mounts for the machine guns and the absence of doors, replaced by chains. The circular bronze badge is also clearly visible.

The Autocarretta 36 P could carry a rifle squad, of commander, six fusiliers and two machine gunners with two Breda 30 light machine guns, plus the driver.

Front and in North Africa. Production of the Autocarretta series ceased in 1942, by which time the new SPA CL 39 was being issued; more than 5,000 of all series were produced.

The Autocarrette 35 and 36 enjoyed some degree of export sales: in 1940 a number of the 36 P version were sold to Portugal, following sales of unspecified numbers of the model 35 to Spain and 36 P to Ecuador; in 1942 100 of the model 36 were sold to Hungary.

Technical Description

The OM series of autocarrette were produced in a number of different series and configurations, mainly marked by changes in body styles. All these were characterised by an open cab mounted over the engine, although a retractable canvas top rolled up behind the driving compartment could be used to cover the top of the compartment.

The Autocarretta was the first four-wheel drive vehicle (except the artillery tractors) to be adopted by the *Regio Esercito*; all four wheels steered. The Autocarretta 35, like the 32 from which it derived, had a wooden body and was intended as a cargo carrier; the model 36 Mt likewise had a wooden cargo body, whereas the 36 P (or 36 DM) had a metal body consisting of three transverse rows of benches that enabled ten passengers (the driver plus an armed rifle squad, of a commander, two machine gunners and six fusiliers) and two Breda 30 light machine guns, or one Breda 37 heavy machine gun, to be carried. Both versions of the model 36 had a retractable canvas top that could be erected over the troop and driver's compartments.

The wheels on the Autocarretta 35 were stamped steel disc wheels with ten lightening holes (five large and five small, alternating) with semi-pneumatic tyres; the wheels on the 36 were more modern-looking stamped steel disc wheels with six lightening holes (four small and two somewhat larger) fitted with pneumatic tyres. The semi-pneumatic tyres could receive

The Autocarretta 36 Mt cargo carrier, without and with protective canvas cover.

traction improving devices made of metal tyre shoe attachments. All four wheels had independent suspension, with semi-elliptical leaf springs. The brakes were drum brakes on all four wheels, with the hand brake acting on the transmission. The electrical system was based on a dynamo, but starting was by means of a hand crank. The engine was a four-cylinder gasoline engine developing 23 HP; it was cooled by forced air (a feature studied for use in very cold climates, like the Alps), each separate cylinder having cooling fins. The transmission had four forward and one reverse speeds and a reduction gear. The autocarrette

The Autocarretta 36 Mt, with Superflex Stella Bianca pneumatic tyres. The rear hatch is shown in the two positions closed and open.

The Autocarretta 37 during factory trials. (OM)

This Autocarretta 37 under test is fitted with semi-pneumatic tyres and transports various wheels with pneumatic tyres and Artiglio treads. External tools are absent. (OM)

were equipped with a backstop device, acting on the primary shaft of the gearbox, useful in uphill starts to avoid the use of the brakes. There was also a mechanical stabilisation system, designed to dampen the vertical oscillations of the front differential.

Variants

On the cargo bed of some Autocarretta 35s a sprinkling system for ground decontamination was installed – the Attrezzatura Irroratrice modello 33. Additionally at the same time a smoke-producing system was manufactured – the Attrezzatura Nebbiogena Modello 33. In summary, this latter consisted of a specially made drum of about 300 litres capacity containing the smoke liquid to be sprayed, a cylinder of compressed-air to provide pressure and a series of pipes with atomising nozzles. Both these special vehicles were assigned to chemical units. Later, for the Autocarretta 32 and Autocarretta 36 a very similar smoke-producing system was made – the Attrezzatura Nebbiogena Modello 39.

On a small but unspecified number, perhaps 20, of chassis of the model 36 an armoured superstructure was mounted, becoming the Autocarretta Ferroviaria Blindata Modello 1942 (armoured railway truck). Some vehicles of this kind were used

The smoke-producing system Attrezzatura Nebbiogena modello 39 could be installed on the Autocarretta 32 and 36. (ACS)

in Dalmatia and Slovenia from December 1942 for the control and safety of narrow gauge (76cm) railway lines, and some were used by the Germans after the Armistice.

Specifications

- Designation: Autocarretta OM 32, 35 and 36
- Producer: Officine Meccaniche (OM), Brescia
- Years produced: 1932–1942
- Number produced: more than 5,000
- Length: 3,770mm (12ft 4in) model 35; 3,910mm (12ft 10in) model 36 Mt; 4,170mm (13ft 8in) model 36 P
- Width: 1,300mm (4ft 3in) model 35; 1,420mm (4ft 8in) model 36 Mt and 36 P
- Height: 2,200mm (7ft 3in) model 35; 2,100mm (6ft 10in) model 36 Mt and 36 P
- Unladen weight: 1,580kg (3,483lbs) model 35; 1,600kg (3,527) model 36 Mt; 1,650kg (3,638) model 36 P
- Carrying capacity: 800kg (1,764lbs) model 35 and model 36 Mt; 10 armed soldiers model 36 P
- Wheelbase: 2,000mm (6ft 7in)
- Front track: 1,100mm (3ft 7in) model 35; 1,070mm (3ft 6in) model 36 Mt and 36 P
- Rear track: 1,100mm (3ft 7in) model 35; 1,070mm (3ft 6in) model 36 Mt and 36 P
- Minimum turning radius: 3,500mm (11ft 6in) model 35; 4,200mm (13ft 9in) model 36 Mt and 36 P
- Minimum clearance: 450mm (1ft 6in)
- Bed, internal length: 1,700mm (5ft 7in)
- Bed, internal width: 1,200mm (3ft 11in)
- Bed, internal height: 500mm (1ft 7in)

Autocarretta Ferroviaria Blindata modello 1942 (armoured railway truck) made on the model 36 chassis.

- Towing capacity: 1,000kg (2,204lbs)
- Tyres: Celerflex 120 x 670 model 35 and model 37; Superflex Artiglio or Stella Bianca 7.00 x 18 model 36 Mt and 36 P
- Engine: AM, four-cylinder in-line, forced air-cooled, 1,616cc, 20 HP @2,400 rpm
- Fuel: Gasoline
- Transmission: four speeds forward, one reverse, with reduction gear
- Fuel capacity: 35 litres (9.2 US gallons, 7.7 Imperial gallons)
- Drive layout: 4x4
- Maximum speed: 25km/h (15.5mph) model 35; 45km/h (28mph) model 36 Mt and 36 P; 36km/h (22.3mph) model 37
- Range (on road): 160km (99 miles)

Autocarro leggero SPA 25 C

The SPA 25C still had the structure of a First World War truck.

Developmental and Service History

The SPA 25 C was developed as a civilian truck, in two versions with different engines: the 25 C/10 and the more powerful 25 C/12.

With the end of production of the Fiat 15 ter light military truck in 1922, the Italian military authorities urgently needed a replacement light truck in the ambulance configuration. The SPA 25 C/10 version was found to be suitable as a replacement, and in 1924 the *Regio Esercito* acquired a first batch of chassis with ambulance bodies fitted by Garavini of Turin. Increasing numbers of the 25 C/10 were then acquired in a number of different body styles. The last series produced had a closed cab with full doors, but without windows, a different radiator and a redesigned bonnet.

A SPA 25 of the last series produced (the photo is from 1940). The bodywork now features a closed cab with full doors but no windows. (ACS)

SPA 25 C/10 ambulance used during the Spanish Civil War (1938). It mounts new disc wheels with pneumatic tyres.

Close-up view of a SPA 25 C/10 ambulance. Note the logistics lettering stencilled in white on the side of the cab.

The 25 C/10 served in various mainly secondary roles throughout the Italian campaigns in East Africa and Spain, and the Second World War. Many examples of SPA 25C were used by the *Regia Aeronautica*. Poland also took delivery of 375 SPA 25 C trucks based on a 1924 order.

Technical Description

The SPA 25 C/10 was a 4x2 truck. The cab of the truck version was an open cab that had a removable canvas top; the ambulance version had a cab with a closed hard top with half-doors and canvas side curtains. Successive versions saw modifications to fender, radiator and bonnet styles. The two-axle truck was a rear-wheel drive vehicle. Semi-pneumatic tyres were mounted on pressed steel disc wheels, dual on the rear axle; later they were replaced by pneumatic tyres. Drive was right-hand. The frame was a ladder type and suspension consisted of semi-elliptic leaf springs; brakes were mechanical drum brakes. Some vehicles had an electrical system with a battery for lights and self-starting, while most apparently used a hand crank for starting. Following common practice for Italian military trucks, acetylene emergency lamps were fitted on the 25 C, in addition to the electric headlights.

The engine was a four-cylinder gasoline engine developing 39 HP and was fitted with a Zenith carburettor. The manual transmission consisted of four forward speeds and one reverse.

The truck bed was wooden with fixed sides and had a hinged tailgate. A canvas top was provided for the bed. There were small storage lockers on the running boards on both sides, and the spare tyre was mounted on a post which in turn was mounted on the frame next to the driver's compartment.

Variants

In addition to the basic cargo body, variants included passenger bus, ambulance, mobile workshop, refrigerated truck and tank truck versions. The truck was also built in a version with a larger engine, designated the SPA 25 C/12, small numbers of which were used by the military.

On the SPA 25 C/10 a smoke generator was mounted consisting of two tanks of 350 litres of liquid, a cylinder of compressed-air and a diffusion system. Although ready in 1935, this vehicle was rarely used, supplanted in the role by the OM 35 and 36 trucks.

Specifications

- Designation: Autocarro Leggero SPA 25 C
- Producer: Società Piemontese Automobili (SPA), Turin
- Years produced: 1925–1934
- Number produced: Not Known
- Length: 5,700mm (18ft 8in)
- Width: 1,800mm (5ft 11in)
- Height: 2,900mm (9ft 6in)
- Unladen weight: 2,300kg (5,071lbs)
- Carrying capacity: 2,500kg (5,512lbs); ambulance version 1,800kg (3,968lbs)
- Wheelbase: 3,250mm (10ft 8in)
- Front track: 1,510mm (4ft)
- Rear track: 1,510mm (4ft)
- Minimum turning radius: 6,750mm (22ft 2in)
- Minimum clearance: 350mm (1ft 2in)
- Bed, internal length: 2,900mm (9ft 6in)
- Bed, internal width: 1,680mm (5ft 6in)
- Bed, internal height: 700mm (2ft 3in)
- Tyres: semi-pneumatic tyres 895 x 135; pneumatic tyres Cord 32 x 6
- Engine: SPA C10 four-cylinder water-cooled, 2,700cc, 39 HP @2,500 rpm
- Fuel: Gasoline
- Transmission: four speeds forward, one reverse
- Fuel capacity: 80 litres (21 US gallons, 17.6 Imperial gallons)
- Drive layout: 4x2
- Speed: 50km/h (31mph)
- Range (on road): 290km (180 miles)

Autocarro Leggero Militare SPA 38 R and 36 R

Developmental and Service History

In early 1933 the *Regio Esercito* asked Fiat to develop a new 2-ton truck to replace the SPA 25 C. Two versions were ordered, one with a liquid-cooled engine and the other with an air-cooled engine. The two prototypes were ready in early 1934 and were subjected to a series of trials. Having met the required specifications, both truck models were adopted in early 1935; production was undertaken by SPA, a Fiat subsidiary.

A factory-new SPA 38 R, complete with canvas top. (Fiat)

Left side, without canvas top. (MSGR)

SPA 38 R trucks ready for delivery. (Fiat)

The SPA 36 R. The differences are evident in the bonnet and radiator grille. (Fiat)

SPA 38 R ambulance and bus civilian configurations. (MSGR)

The truck with liquid-cooled engine was named 38 R and was more widely issued. The air-cooled 36 R, was instead envisioned for use in desert and mountain environments; it was produced in limited numbers because field use in Italian East Africa in 1937 and in Libya in 1938 revealed that it was subject to frequent breakdowns, highlighting poor overall reliability. The 38 R, on the other hand, had participated with the Italian contingent in the Spanish Civil War and proved to be a solid truck with good performance on roads, although somewhat less so cross-country. In Spain it was used to tow infantry 65/17 guns, as well as acting as a prime mover for field and heavy field artillery. It saw extensive use in the Second World War in a variety of roles.

SPA 38 R military ambulance.

A SPA 38 R belonging to the Italian *Corpo Truppe Volontarie* in Spain. The camouflage, that extends to the canvas cover and licence plate type, is also visible on other vehicles featured in this study. (AUSSME).

The 38 R, whose configuration was similar to that of most Italian military trucks of the early 1930s, was a simple, robust machine that was easy to drive and performed well over rough ground. These qualities induced the French to order 500 examples for the *Armée de l'Air* (the French Air Force). About 400 vehicles were delivered before the war; the French used them as prime movers for the 75mm gun. Throughout the war, the truck was used in large numbers by the *Regio Esercito*, both in the standard version as well as the so-called colonial version, which had an oil-bath air filter, battery and a new electric starter motor, a 100-litre fuel tank, power take-off and other features. The *Regia Aeronautica* adopted a slightly dissimilar version of the truck, the model 38 RA, with a wider wheelbase (3,600mm), different rear brakes and other minor changes; an ambulance version based on the 38 RA was also used by the *Regia Aeronautica*. During the post-war period, production resumed of the 38 R/45, which had a battery, an 88-litre fuel tank, and a new completely enclosed cab. The 38 R remained in service in the Italian Army for many years after the end of the war.

Technical Description

The 38 R had a conventional 4x2 layout with a standard cab configuration. It was a two-axle truck with rear-wheel drive and dual rear wheels. The cab doors were half-doors; canvas side panels were provided to close off the cab in inclement weather. As with most Italian military trucks of the period, it had right-hand drive. The ladder frame consisted of two steel side frames and seven cross-members. The suspension consisted of leaf springs.

A SPA 38 R with a Red Cross flag stationary in the desert of Sidi Barrani in September 1940. (ACS)

An example bogged down in the Ukrainian mud; the soldiers attempting to push the truck have spread branches on the ground in an attempt to provide some traction. (AUSSME)

The engine was the Fiat 18 R which was also used on the Fiat Dovunque truck. The 18 R was a four-stroke gasoline engine assisted by a Weber 42 AK carburettor. The engine was mounted on guide rails that allowed it to slide forward for repairs and inspection. Ignition was by a Marelli FL4 magneto. The 38 R Coloniale had instead a Marelli A 20/12 electric starter and two packs of Marelli 3MF15 6-volt batteries. Cooling was by a centrifugal pump and a fan; the radiator consisted of eight elements

A SPA 38 R carrying a contingent of Bersaglieri in North Africa. (AUSSME)

Two 38 Rs equipped with Breda 20mm machine guns escort a column led by a Fiat 626. Eastern Front, summer 1941. (ACS)

which could be closed off from each other. The manual transmission, which was not synchronised, had four forward and one reverse gears. Drum brakes were present on all wheels. The truck bed was wood with fixed sides and had a hinged tailgate with mounting steps. Five wooden benches could be fitted to the bed to carry troops. A canvas top with five bows was fitted. The spare tyres were stored below the body, under storage lockers.

This accident involving a SPA 38 allows the chassis to be seen from below.

The mobile repair shop Modello 37 on the SPA 38 R chassis. Viberti designed a similar model.

Variants

In addition to the basic truck, variants included a refrigerated truck, ambulance, mobile repair shop, field office, passenger bus, and at least one mobile library.

The 20mm Breda anti-aircraft cannon was often mounted on the bed of the 38 R.

A SPA 36 R, with its characteristic oval radiator grille, precedes a 38 R.

Specifications

- Designation: Autocarro Leggero Militare SPA 38 R/SPA 36 R
- Producer: Società Piemontese Automobili (SPA), Turin
- Years produced: 1935–post-war
- Number produced: Not Known
- Length: 5,783mm (18ft 11in) 38 R; 5,830mm 36 R (19ft 2in)
- Width: 2,070mm (6ft 10in) 38 R; 2,000mm (6ft 7in) 36 R
- Height (with canvas cover): 2,780mm (9ft 2in)
- Unladen weight: 3,200kg 38 R; 3,250kg 36 R
- Carrying capacity: 2,500kg (5,512lbs)
- Wheelbase: 3,500mm (11ft 6in)
- Front track: 1,545mm (5ft)
- Rear track: 1,427mm (4ft 8in)
- Minimum turning radius: 5,700mm (18ft 8in)
- Minimum clearance: 250mm (10ft)
- Fording depth: 600mm (2ft)
- Bed, internal length: 3,200mm (10ft 6in)
- Bed, internal width: 1,980mm (6ft 6in)
- Bed, internal height: 670mm (2ft 2in)
- Tyres: Cord 32 x 6; Ultraflex 210 x 20
- Engine: Fiat 18 R, 4,053cc, water-cooled, 56 HP @2,000 rpm 38 R; Fiat six-cylinder air-cooled, 4,426cc, 50 HP @2,000 rpm 36 R
- Fuel: Gasoline
- Transmission: four speeds forward, one reverse
- Fuel capacity: 38 R, 108 litres (28.5 US gallons, 23.8 Imperial gallons); 36 R, 100 litres (26.4 US gallons, 22 Imperial gallons)

LIGHT TRUCKS 115

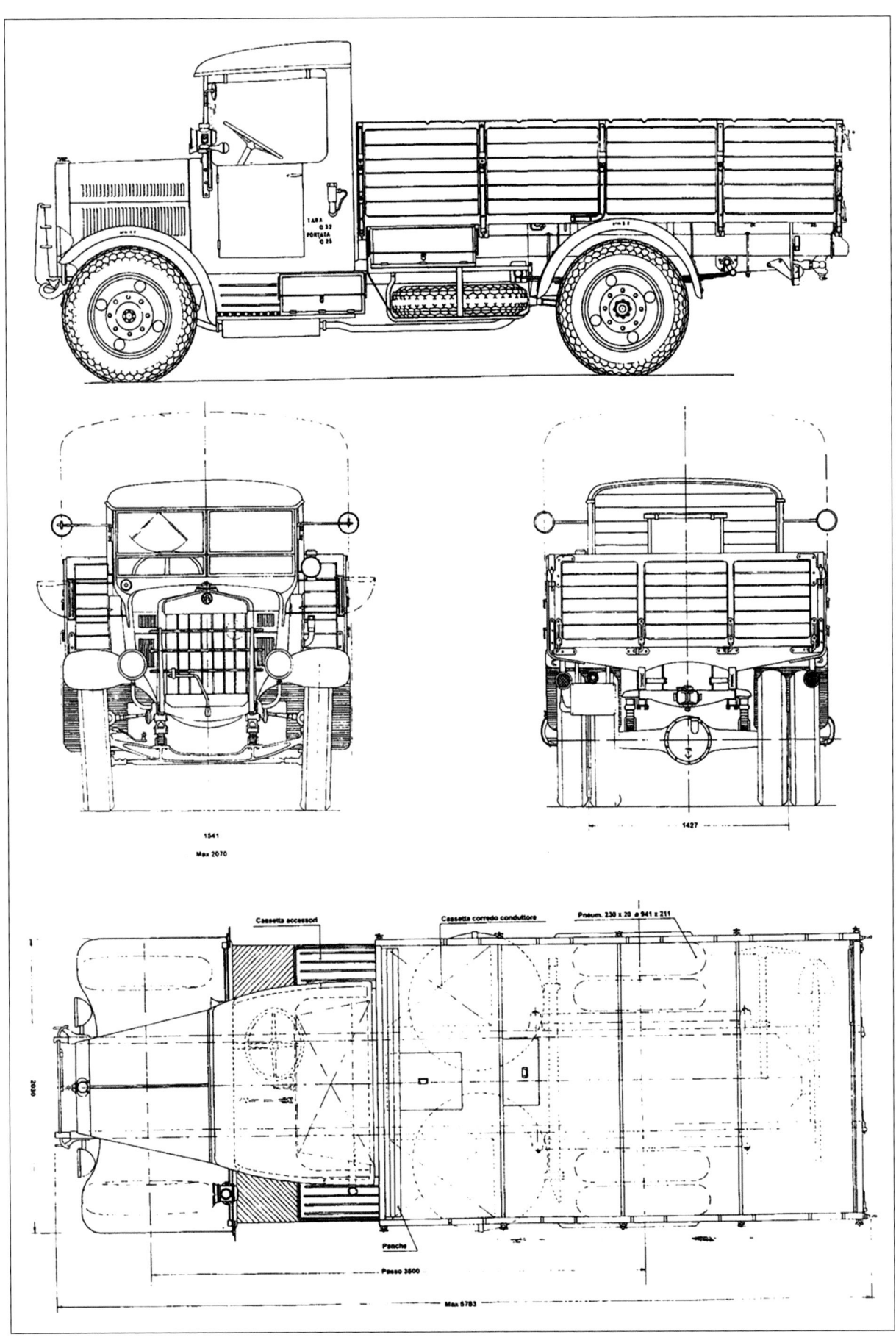

Line drawing of the SPA 38 R

- Drive layout: 4x2
- Speed: 52km/h (32mph)
- Range (on road): 310km (192.5 miles)
- Range (cross-country): 290km (180 miles)

Autocarro Leggero Fiat 618 MC

The Fiat 618 Militare Coloniale light truck. (Fiat)

A Fiat 618 MC in East Africa.

Men of the *6ª Divisione Camicie Nere 'Tevere'* of the MVSN, in Somalia in the second half of the thirties.

A Fiat 618 of the Italian CTV corps in Spain.

Fiat 618, possibly destroyed during the battle of Santander in Spain, in August 1937. It belonged to *724th Bandera Inflessibile* of the *7th Gruppo Banderas*, part of the *2nd Fiamme Nere* Volunteer Division of the MVSN. Each *Bandera* of the MVSN corresponded to a battalion and the *Gruppo Banderas* to a regiment.

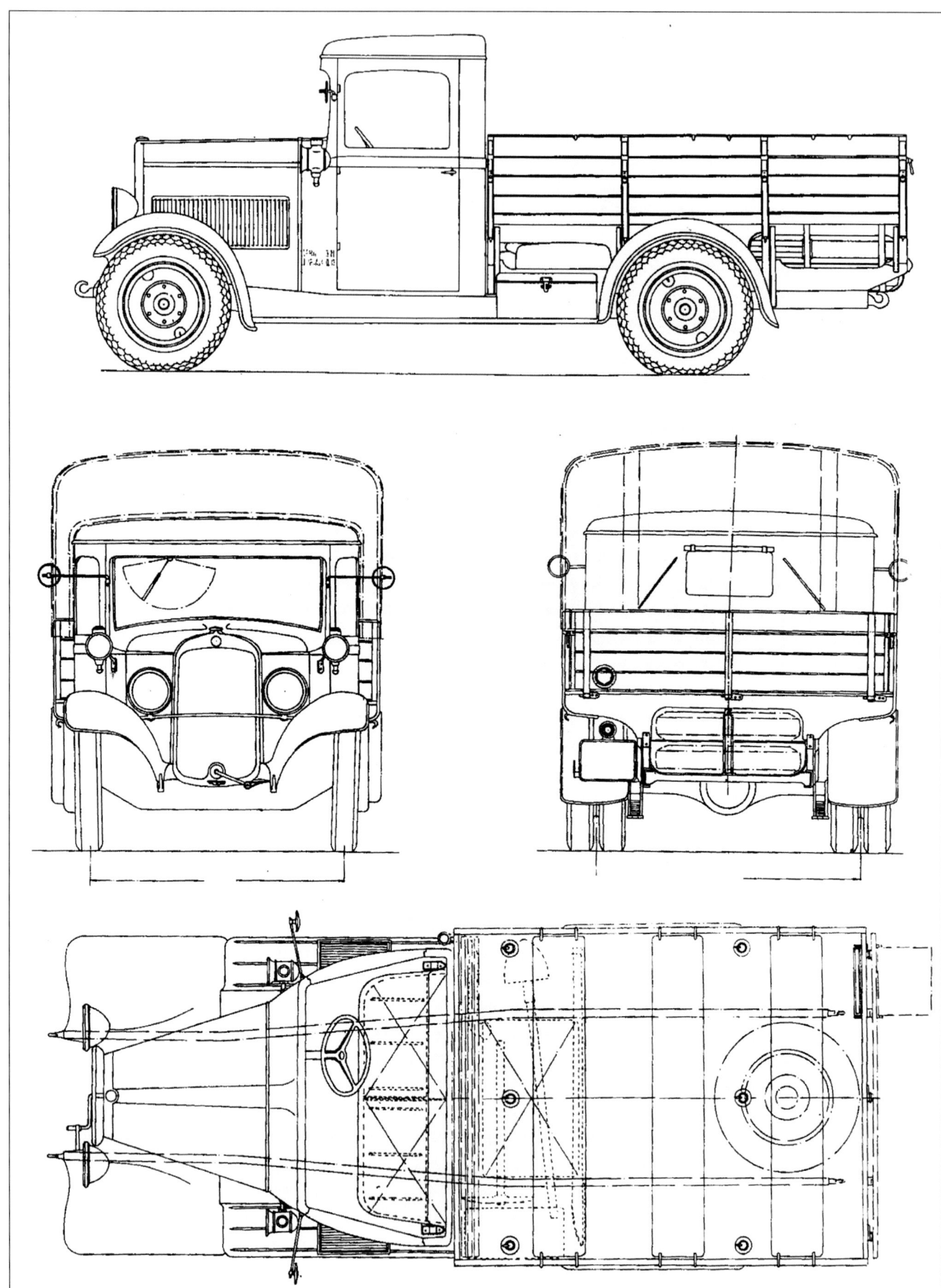

Line drawing of the Fiat 618, August 1935. (Fiat)

The ambulance bodywork made by Viberti.

Developmental and Service History

The Fiat 618 Militare Coloniale, more commonly referred to as simply the 618 MC, was the military version of the civilian Fiat 618, designed for light transportation duties that had entered production in 1934. More precisely, it was derived from modified version 618 C (Coloniale), which entered service in 1935 and was conceived for use in Italian North and East Africa. In these territories, it proved to have several problems, including engine overheating, a weak transmission, problematic suspension and limited cargo capacity when operating in the mountains. Eventually, despite its widespread use, the truck was relegated to duties as a squad carrier for infantry units.

Nevertheless, the 618 MC saw extensive service in Ethiopia and Somalia in 1935–1936 (probably more than 3,000 examples), and later almost 1,700–2,000 were assigned to the Italian *Corpo Truppe Volontarie* (the CTV, the Italian Volunteer Corps) that served in the Spanish Civil War. In mid-1937 each 37mm anti-tank battery, equipped with the German 3.7cm Pak 36 gun, in the CTV in Spain was issued with six Fiat 618 MC trucks as prime movers for the guns. Later that year, a number of 618 MC trucks were fitted with St Etienne machine guns for anti-aircraft defence duties. The next year, following exercises carried out in Libya, the truck was once again judged not sturdy and was thus considered suitable only for light duties.

The Fiat 618 MC was used predominantly by the *Regio Esercito*, but other users included the *Milizia Volontaria per la Sicurezza Nazionale* (MVSN), the Finance Guard, the Public Security Corps and the *Regia Aeronautica*. The Fiat 618 was manufactured under licence in Poland for both the civilian and the military market. Although its production in Italy was terminated in 1937, being replaced by the CL 39, the 618 MC continued to serve throughout the war in lieu of more suitable vehicles.

Technical Description

The Fiat 618 MC was a conventionally laid out 4x2 truck with rear-wheel drive and dual rear wheels. The closed cab with full doors had a wooden frame with iron reinforcements, covered with sheet metal. The windshield was in one piece and could be opened forward. A manual windshield wiper on the right side, the driver's side, and rear-view mirrors were optional accessories. The cargo body was made of wood, with a one-piece tailgate. Soldiers on board could sit on three removable benches. Pneumatic tyres were mounted on pressed steel wheels that had two lightening holes per wheel; two spare wheels and tyres were carried under the rear of the cargo bed. Brakes were hydraulic drums on all four wheels. The hand parking brake acted on the transmission. The electric lighting system used a dynamo and no battery was present; the engine was started by means of a hand crank. The engine was a four-cylinder gasoline engine which developed 43 HP; the transmission had four forward and one reverse gear.

Variants

Variants of the 618 MC included a radio van, ambulance and passenger bus.

Specifications

- Designation: Autocarro Leggero Fiat 618 Militare Coloniale
- Producer: Fabbrica Italiana Automobili Turin (Fiat), Turin
- Years produced: 1934–1937
- Number produced: Not Known
- Length: 4,700mm (15ft 5in)

- Width: 1,940mm (6ft 4in)
- Height: 2,500mm (8ft 3in)
- Unladen weight: 2,115kg (4,662lbs)
- Carrying capacity: 1,200kg (2,645lbs)
- Wheelbase: 3,050mm (10ft)
- Front track: 1,490mm (4ft 11in)
- Rear track: 1,540mm (5ft)
- Minimum turning radius: 5,750mm (18ft 11in)
- Minimum clearance: 200mm (8in)
- Tyres: Superflex 6.00 x 20
- Engine: Fiat model 118A, four-cylinder in-line water-cooled, 1,944cc, 43 HP @3,800 rpm
- Fuel: Gasoline
- Transmission: four speeds forward, one reverse
- Fuel capacity: 60 litres (15.8 US gallons, 13.2 Imperial gallons)
- Drive layout: 4x2
- Maximum speed: 65km/h (40mph)

Fiat 508 M Camioncino and Furgoncino

The Fiat 508 Camioncino three-speed, left-hand drive.

The Fiat 508 Furgoncino three-speed, right-hand drive.

Developmental and Service History

In 1932 the *Regio Esercito* adopted a compact car, the Fiat 508 M, which proved to be well-liked and successful (see above). Subsequently, a small pick-up truck referred to in Italian as a 508 Camioncino was developed based on the 508 A and 508 B chassis (three-speed and four-speed respectively), as was a small, enclosed van, the 508 Furgoncino, for both the civil and military markets.

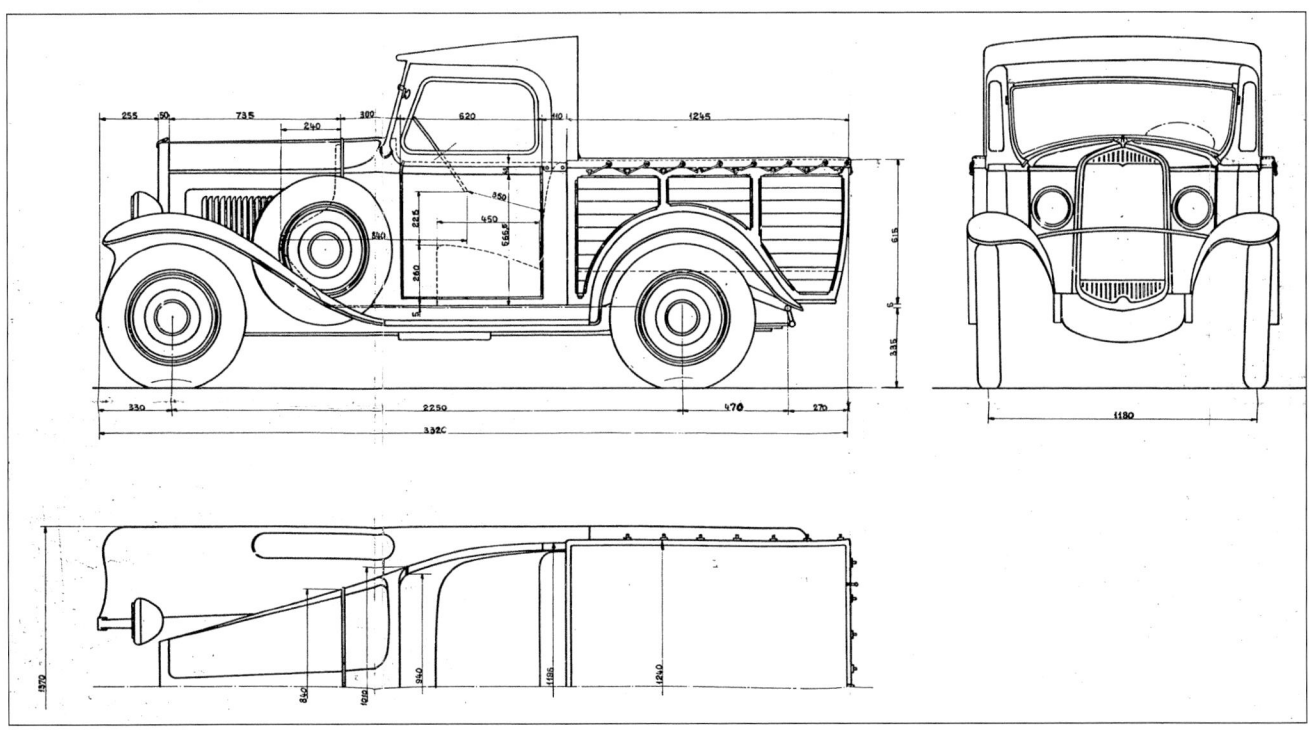

Line drawing of the 508 Camioncino three-speed, left-hand drive, December 1931. (Fiat)

The Fiat 508 Camioncino four-speed, recognisable by the different shape of the grille and the engine hood.

Two Camioncini four-speed trucks in Asmara, Eritrea, in the second half of the 1930s.

A 508 Camioncino four-speed, without spare wheel, in Asmara in 1943. (V. Cerenzia / www.acrinews.it)

The civilian Fiat 508 Camioncino four-speed, with a slightly different body.

These small trucks served the *Regio Esercito* throughout the war, and were also used by the *Regia Marina* and the *Regia Aeronautica*, seeing heavy use (at least 497 Camioncini and 20 Furgoncini) in Italian East Africa.

Technical Description

The 508 Camioncino and Furgoncino shared almost all of their mechanical components with the 508 A/B/M passenger car on which they were based. The Camioncino had a metal cab with half-doors, a canvas top and rear panel. The Furgoncino had an all-metal cab and body and had full doors. The windshield was of a one-piece type. The bed, bed sides and tailgate of the Camioncino were made of wood, with folding benches for transporting troops, while the body of the Furgoncino, as noted, was all metal. Pneumatic low pressure tyres were mounted on pressed steel wheels and there were two spare wheels/tyres mounted on the sides of the bonnet. All four wheels had drum brakes. Suspension was leaf springs front and rear with enhanced hydraulic

The Fiat 508 Furgoncino four-speed. (Fiat, ACS)

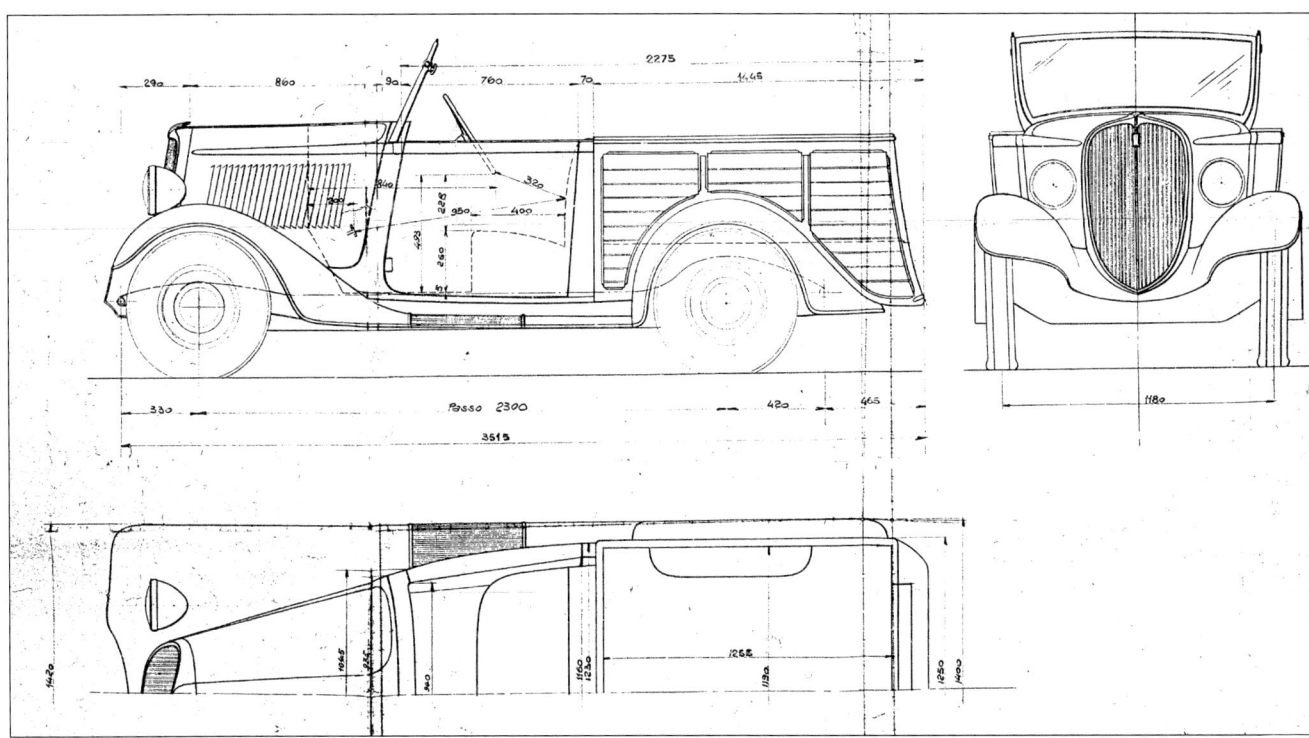

Bodywork design for a 'Camioncino 508 III' dated 20 September 1934. The vehicle is based on the four-speed 508 B, but with some elements (mudguards and cargo bed with a 'protruding tail') that anticipate the future 508 L Camioncino. (Fiat)

shock absorbers compared to civilian models. The engine was a four-cylinder gasoline engine developing 20 HP. The manual transmission of the first version had three forward and one reverse gears, while the second version and variants had four forward and one reverse gears. The military version could have an oil air filter instead of a dry one.

Variants

On a few dozen examples of Camioncino (50 or 68, depending on the sources), belonging to 508 M and 508 C / 1100 version, a weapon system of two coupled 8mm Fiat 35 machine guns was mounted.

On the long wheelbase Furgoncino a small ambulance could be set up, the body style differing depending on the coachbuilders.

Specifications

- Designation: Fiat 508 M Camioncino; 508 M Furgoncino
- Producer: Fabbrica Italiana Automobili Turin (Fiat), Turin
- Years produced: 1933–1937
- Number produced: NA
- Length: 3,320mm (10ft 11in) first series; 3,515mm (11ft 6in) second series
- Width: 1,380mm (4ft 8in)
- Height: 1,450mm (4ft 9in) with top down; 1,600mm (5ft 3in) with top up (and pick-up version)
- Unladen weight: 715kg (1,576lbs)
- Carrying capacity: 350kg (772lbs) Camioncino; 300kg (661lbs) Furgoncino
- Wheelbase: 2,250mm (7ft 4in) first series; 2,300mm (7ft 6in) second series
- Front track: 1,800mm (5ft 9in)
- Rear track: 1,800mm (5ft 9in)
- Minimum turning radius: 4,750mm (15ft 7in)
- Minimum clearance: 220mm (9in)
- Tyres: Superflex 4.25 x 17
- Engine: Fiat 108 M four-cylinder, side-valve, water-cooled, 995cc, 20 HP @3,400 rpm first series (three-speed); 24 HP @3,800 rpm second series (four-speed)
- Fuel: Gasoline
- Transmission: three speeds forward, one reverse on first series; four speeds forward, one reverse (3rd and 4th gears synchronised) on second series
- Fuel capacity: 26 litres (6.9 US gallons, 5.7 Imperial gallons)

- Drive layout: 4x2
- Maximum speed: 65km/h (40mph)
- Range (on road): 310km (193 miles)

Fiat 508 C and 1100 Camioncino and Furgoncino

The Fiat 508 C Camioncino or 'Musetto' (small snout). (CSM)

The driver of a camouflaged 508 C Camioncino with a *Regio Esercito* number plate talking to a *Milizia della Strada* officer. (ACS)

Fiat 508 C Camioncino and Lancia 3Ro trucks in North Africa.

Budapest, 1941. A convoy with several Bianchi Miles trucks ready to head to the Eastern Front. In the foreground, a Fiat 508 C Camioncino with the symbol of the *Stato Maggiore Regio Esercito* (Army Staff). (ACS)

508 C Camioncini from the *Stato Maggiore Esercito* on the Eastern Front in 1941. (ACS)

This image is part of a series of poses taken on Agfa colour film by a German soldier during the transfer trip and stay in North Africa. On the left, an Opel Olympia 38 that appears camouflaged in the *Tropen* colours employed by the Wehrmacht since March 1942. On the right, a Fiat 508 C Camioncino belonging to the *Regio Esercito* whose grey-green colour is only partially covered by the *kaki sahariano* (sand-yellow) applied by spray gun.

The Fiat 1100 Camioncino or *Musone* (big snout). (Fiat)

A 1100 Camioncino used by the *Ufficio Propaganda Regio Esercito, Servizio Assistenza Truppe* (Army Propaganda Office, Troops Assistance Service).

Developmental and Service History

In the second half of the 1930s, the Italian armed forces incorporated a military version of the popular civilian car Fiat Nuova Balilla sedan, i.e. the third series of the Fiat 508, which was called the 508 C Militare (see above). As already done for the previous models, on its base a small pick-up truck version (*camioncino*) and an enclosed van (*furgoncino*) with an all-metal body were developed. The evolution of these two vehicles began from the exit from the market of the equivalents based on the old 508 M (see the Fiat 508 M Camioncino and the Fiat 508 M Furgoncino).

1100 Camioncino (*Musone*) crossing a pontoon bridge over the Donets, a river which traverses both Russia and the Ukraine. Its light colour, perhaps *kaki sahariano*, is interesting compared to the dark grey-green of the other vehicles. (ACS)

The 508 L Camioncino (small snout, long frame) was produced for the civilian market but also used by the *Regio Esercito*.

A 508 L Camioncino, painted grey-green.

The first series of the new pick-up truck was manufactured in two variants: the Fiat 508 C Camioncino (in this case, C stands for *corto*, i.e. short wheelbase) was produced until the end of the Second World War for military use only; the 508 L version (*lungo*, long wheelbase) was both civilian and military. The same was true for the 508 Furgoncino. Both of these small vehicles were adopted in 1939 and saw service throughout the war in all three branches of the Italian armed forces.

From 1941, both the Camioncino and the Furgoncino were produced on the basis of the restyled Nuova Balilla Fiat 508 C, better known as the Millecento or '1100'. As with the previous truck, the short wheelbase 1100 C variant was intended for military use only, while the 1100 L long wheelbase variant could be both civil or military.

Unusual rear view of a 508 L Camioncino.

A 1100 L Camioncino (big snout, long frame) captured by Commonwealth troops.

The *Istituto Luce* included a *Reparto Guerra* (War Detachment), with teams of cinematographers and photographers located with the various armed forces. Here a 1100 L Furgoncino in the spring of 1942. The vehicle has a civilian number plate. (ACS)

Technical Description

The 508 C / 1100 Camioncino was a small 4x2 pick-up truck. The body of the military variant was of mixed metal and wood construction: the bonnet, fenders, cowling, and cab roof were of metal, doors were a mix of metal and wood panels (long frame versions had all-metal doors and different mudguards), and the cargo body was wood. Compared to the sedan, the Camioncino frame was strengthened and the tyres mounted were larger. The earlier version based on the Nuova Balilla 508 C model was afterwards christened Musetto (small snout); this because the later version based on the new 1100 was nicknamed Musone (big snout), having a more prominent grille and a slightly different and less slanted bonnet.

In emergencies, these vans were used for other functions, as in this case for the transport of a wounded soldier on the Eastern Front in the summer of 1942. (ACS)

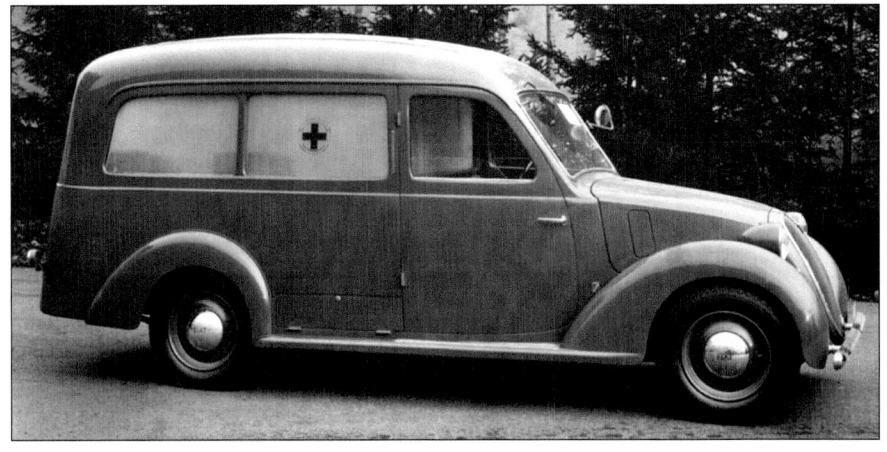

A civilian ambulance based on the 508 L frame. (Croce Verde Lugano)

The enclosed van version of the 508 C / 1100 Camioncino was known as the Furgone (van) or more commonly Furgoncino (small van); the body was all metal, similar to that on many small civilian delivery vans of the era. Because bodies were built by many independent coach builders there could be minor differences in body style and configuration. The independent front suspension on both types used coil springs and hydraulic shock absorbers; the rear suspension consisted of leaf springs, hydraulic shock absorbers and a stabiliser bar. Brakes were hydraulic drum brakes, plus a mechanical hand brake on the transmission. The engine was the same four-cylinder valve-in-head gasoline engine used on the 508 C sedan, developing 30 HP.

Variants

The Camioncino was built in a mobile repair van version. Some 508 M (the initial model), 508 C and 1100 Camioncino were also converted to armed versions, with armament consisting of twin Fiat 8mm model 35 air-cooled machine guns fed by box magazines, mounted on a modified 20mm gun mount.

Two types of ambulance body, one all-metal and the other of wood, were built on the lengthened frame of the Furgoncino of all models.

Specifications

- Designation: Fiat 508 Camioncino Militare; Fiat 508 Furgoncino Militare
- Producer: Fabbrica Italiana Automobili Turin (Fiat), Turin
- Years produced: 1939–1945 ca.
- Number produced: a few thousand Camioncini (508 M and 508 C / 1100 version) until 1943, 1,677 from September 1943 to 1945
- Length: 4,100mm (13ft 5in) 508 C Musetto; 4,115mm (13ft 6in) 1100 Musone

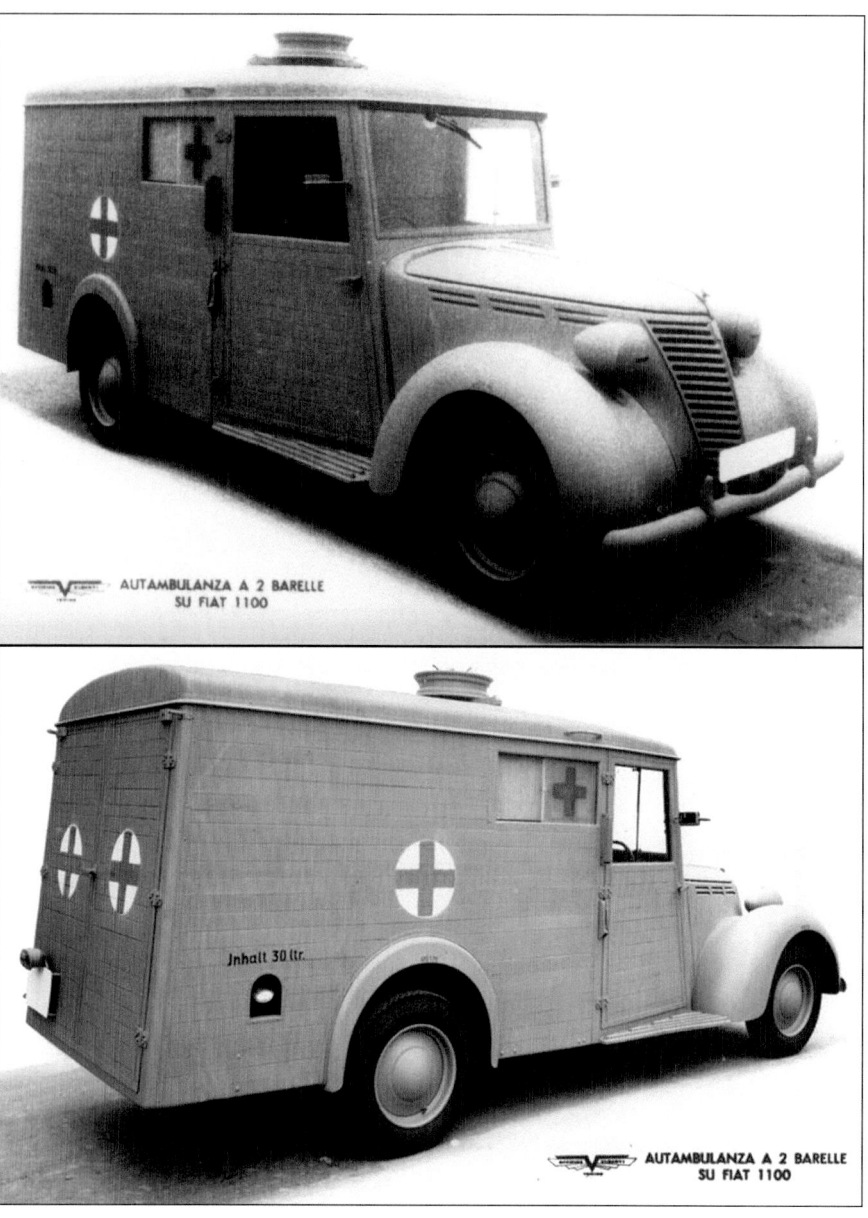

The Fiat 1100 two-stretcher ambulance with wooden bodywork. On the cab top a spare wheel without tyre is fixed. After the Armistice, the vehicle was produced for the Wehrmacht, as evidenced by the German wording next to the tank filler neck (*Inhalt 30 ltr.*, content 30 litres). (Viberti)

An example employed by the Germans, In this case the spare wheel is fitted with a tyre.

- Width: 1,520mm (5ft) 508 C Musetto; 1,660mm (5ft 6in) 1100 Musone
- Height without tarpaulin: 1,680mm (5ft 6in)
- Height with tarpaulin: 2,050mm (6ft 8in) 508 C Musetto; 1,530mm (5ft) 1100 Musone
- Unladen weight: 880kg (1,940lbs) 508 C Musetto; 1,008kg (2,222lbs) 1100 Musone
- Carrying capacity: 420kg (926lbs)
- Wheelbase: 2,427mm (7ft 11in) for *corto*; 2,707mm (8ft 10in) for *lungo*
- Front track: 1,265mm (4ft 2in) for *corto*; 1,316mm (4ft 4in) for *lungo*
- Rear track: 1,254mm (4ft 1in) for *corto*; 1,367mm (4ft 6in) for *lungo*
- Minimum turning radius: 5,000mm (16ft 5in)
- Minimum clearance: 230mm (9in)
- Bed, internal length: 1,540mm (5ft)
- Bed, internal width: 1,600mm (5ft 3in)
- Bed, internal height: 600mm (2ft)
- Tyres: Aerflex 6.00 x 16
- Engine: Fiat 108 C, four-cylinder water-cooled, 1,089cc, 30 HP @4,400 rpm
- Fuel: Gasoline
- Transmission: four speeds forward, one reverse
- Fuel capacity: 30 litres (8 US gallons, 6.6 Imperial gallons)
- Drive layout: 4x2
- Maximum speed: 75km/h (47mph)
- Range (on road): 270km (168 miles)

Autocarro Leggero SPA CL 39

Developmental and Service History

The CL 39 (initially designated the L39) was known or referred to by different names or designations: Autocarretta SPA (to distinguish it from the Autocarretta OM); the *Carro Leggero per Fanteria* (CLF, light infantry truck); Autocarro Leggero mod. 39, but the most commonly accepted designation is the Autocarro SPA CL 39.

The SPA CL 39 had its genesis in the requirement for a simple light infantry truck that could operate in Italy's mountain environment. It had to be less complicated and quicker to manufacture than the OM light trucks, as well as less expensive. The CLF, prototype of the CL 39, was ready in late 1938 and was adopted with some changes in 1939; specifications originally called for two versions of the CL, one with a water-cooled engine and one with an air-cooled engine, but the air-cooled option was discarded. A colonial version of the CL 39 followed in early 1941; it was characterised, in addition to the oil-bath filter, by larger tyres and the cargo bed being lower on the chassis. In addition to acting as an infantry squad carrier, the CL 39, which had good towing qualities, was used to tow a variety of weapons, from the 20mm Breda, to the ubiquitous 47/32 anti-tank gun, and even the larger 75/18 howitzer.

The CL 39 was produced in large numbers and in addition to its use primarily by the *Regio Esercito*, was also used by the *Regia Aeronautica*. In army use, it was assigned on a scale of six CL 39s to each 75/18 artillery group, eight machines to each 81mm

The prototype of the CLF 39 with obsolete, stamped 'artillery wheels' and Celerflex semi-pneumatic tyres. (F. Cappellano)

The SPA CL 39 with Celerflex semi-pneumatic tyres and Dayton-type spoked wheels.

The CL 39 with Superflex Artiglio tyres.

mortar company and four to each 47mm cannon company of the infantry divisions. Additionally, the artillery headquarters (regiment, group, battery) of the armoured divisions *Littorio* and *Ariete* each had three CL 39s. Other examples were assigned to the motorised divisions, both in standard and colonial versions. In 1943, 145 CL 39s were authorised for each infantry division.

Overall, the CL 39 was a very satisfactory and successful vehicle. It was easy to drive, easy to maintain, had excellent visibility, a tight turning radius, good suspension and good fuel consumption.

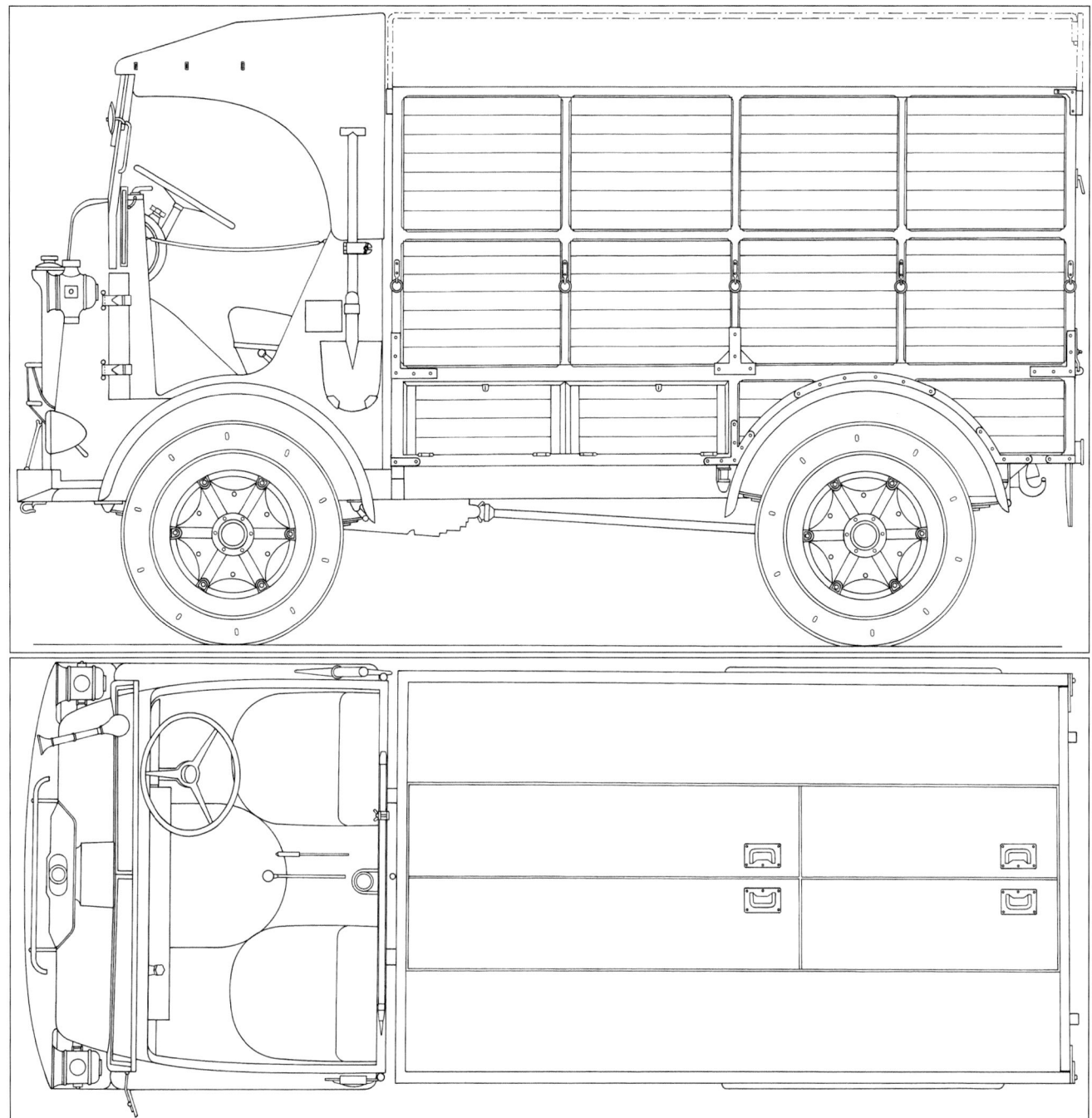

The CL 39 light truck with semi-pneumatic tyres. (Drawing by A. M. Feller – GMT)

The CL 39 saw service on all fronts in which the Italians were engaged, except for Italian East Africa. Following the 8 September Armistice, 198 CL 39s were delivered to German forces in 1944. Following the war, the CL 39 remained in service with the Italian Army until the 1950s.

Technical Description

The CL 39 had been designed as a very simple no-frills machine. It was a 4x2 truck with rear-wheel drive and a cab-over-engine configuration, fitted with the usual, for Italian trucks, right-hand drive. The ladder frame consisted of two rails with five transverse cross-members; a towing pintle was attached to the rearmost cross-member. The cab had a removable canvas top and canvas rear panel. Chain stitches were used in place of the doors. The wooden body had bench seats that could hold eight soldiers, folding back the benches allowed a 1,000kg payload to be carried. There were four storage lockers, two on each side, mounted below the bed. The bed of the metropolitan version of the CL 39 had high sides, whereas the sides on the colonial version were somewhat lower.

The wheels and tyres mounted on the CL 39 underwent an evolution: the earliest version had cast steel wheels with eight spokes and was fitted with Celerflex semi-pneumatic tyres; later versions had six-spoke wheels with 178mm wide Superflex

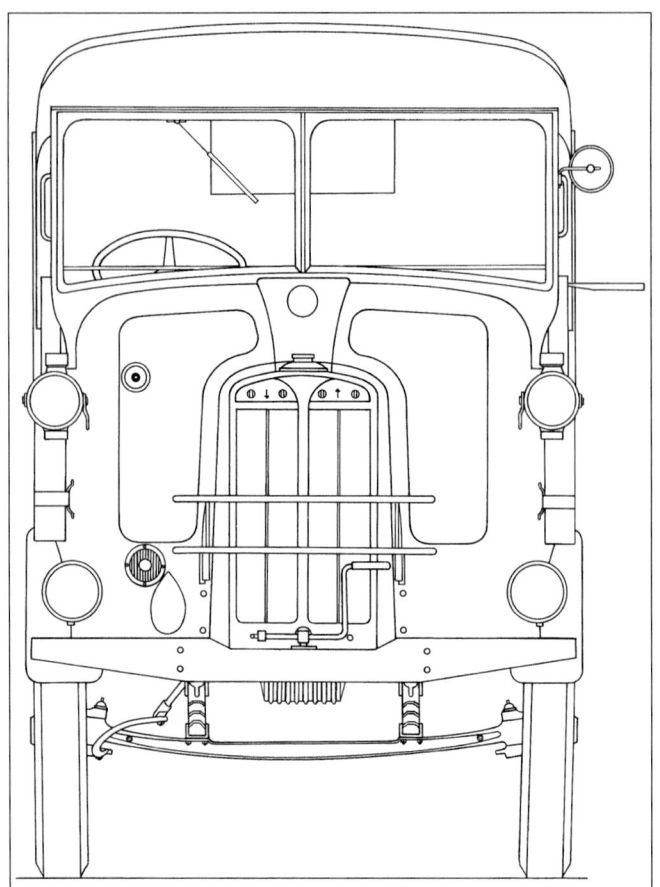
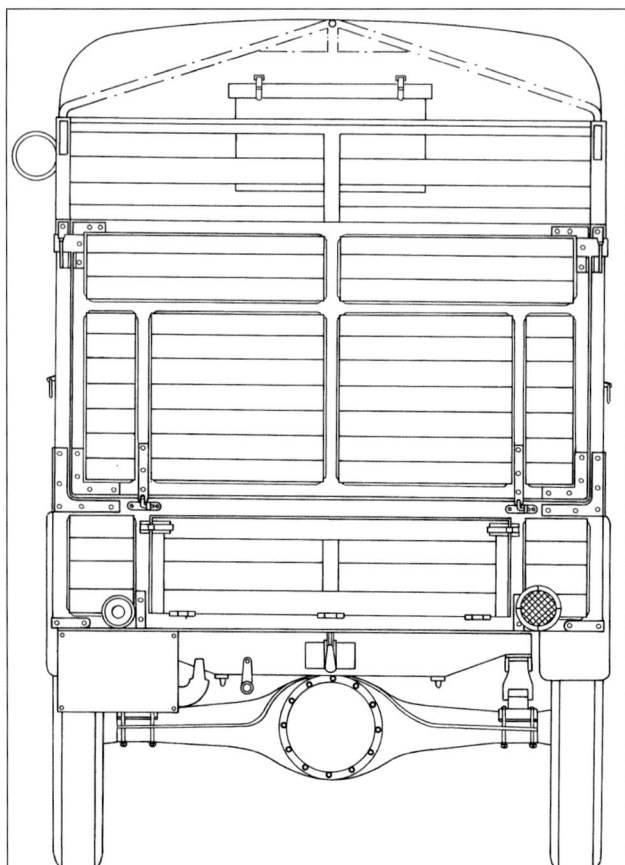

The CL 39 light truck with semi-pneumatic tyres. (Drawing by A. M. Feller – GMT)

The CL 39 Coloniale without and with canvas cover. The tyres are Ultraflex Sigillo Verde 210 x 18 balloon tyres.

Artiglio pneumatic tyres, except the colonial version which mounted the Ultraflex Sigillo Verde very low pressure tyres 210mm wide. The spare tyre on the metropolitan version was mounted beneath the rear of the bed, while the colonial version had the spare tyre mounted behind the cab. The oil-bath air filter of the colonial version was later extended to all versions. Suspension consisted of semi-elliptical leaf springs and shock absorbers. Brakes were hydraulic on all four wheels, while the hand brake acted on the transmission. The electrical system worked on a dynamo. The gasoline engine was a four-cylinder that developed 24 HP. The manual transmission, which was not synchronised, had five forward and one reverse gears; a reduction gear doubled the available speeds. The CL 39 also had a backstop device.

Variants

The CL 39 was built in a standard (metropolitan) version as well as a colonial version (CL 39 C), but changes throughout the production run often blurred the distinction between the two types.

A shower truck (Autobagno) was also manufactured, for decontaminating soldiers exposed to chemicals or gas, but finally used for personal hygiene.

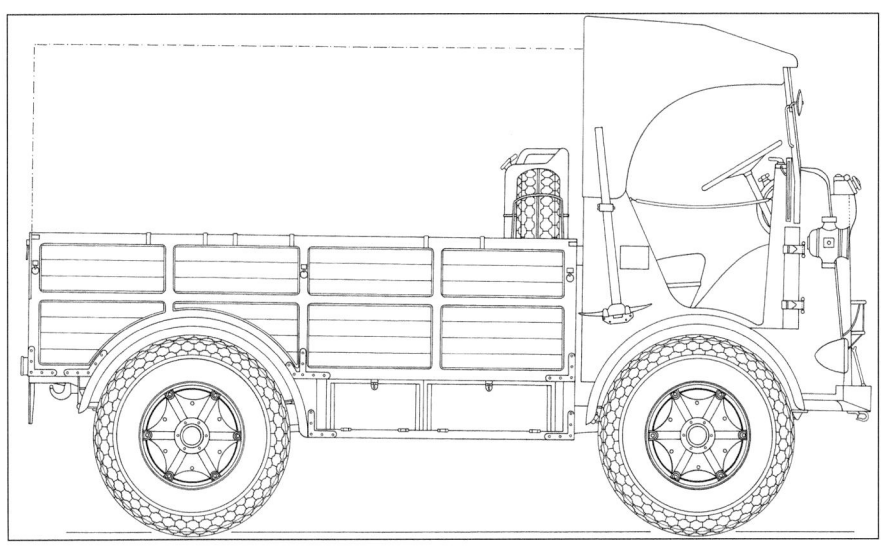

The CL 39 Coloniale. (Drawing by A. M. Feller – GMT)

A CL 39 from the *Regia Aeronautica* in North Africa in 1941. (Bortolotti)

A column of CL 39 light trucks in the Ukraine in 1941. The vehicles carry the regulation bronze vehicle badge and the number plate painted on the front bumper. (ACS)

Struggling with the ford of a watercourse, still in Ukraine. (ACS)

A CL 39 with semi-pneumatic tyres having just crossed a bridge built by Italian engineers over the Dnieper River; note the camouflage netting and the anti-aircraft weapon mounted in the bed. (ACS)

Italian soldiers in North Africa.

A CL 39 of the Italian *Regia Aeronautica* in Albania in winter 1940–1941. Note the different format of the number plate compared to that of the *Regio Esercito* seen earlier, and the absence of the circular bronze badge (ACS)

CL 39 light trucks of the *Giovani Fascisti* (the Young Fascists or GG.FF.) in Libya in the summer of 1941. The first vehicle is equipped with a Fiat 35 machine gun. Some clothing is used to protect the tyres from the heat of the sun. (Museo del Reggimento GG.FF.)

CL 39 shower truck.

Specifications

- Designation: Autocarro Leggero SPA CL 39
- Producer: Società Piemontese Automobili (SPA), Turin
- Years produced: 1939 to mid–1950s
- Number produced: 5,840
- Length: 3,890mm (12ft 9in)
- Width: 1,520mm (5ft)
- Height 2,300mm (7ft 6in)
- Unladen weight: 1,630kg (3,594lbs)
- Carrying capacity: 1,000kg (2,204lbs)
- Wheelbase: 2,300mm (7ft 6in)
- Front track: 1,300mm (4ft 3in)
- Rear track: 1,320mm (4ft 4in)
- Minimum turning radius: 5,000mm (16ft 5in)

Rare image of a CL 39 equipped with improvised armour used in Montenegro in September 1942; it probably belonged to the *XVIII Battaglione Mortai da 81mm* (a mortar battalion) of *18ª Divisione di Fanteria, 'Messina'*.

- Minimum clearance: 245mm (10in)
- Fording depth: 700mm (2ft 3in)
- Bed, external length: 2,480mm (8ft 2in)
- Bed, external width: 1,460mm (4ft 9in)
- Bed, external height: 995mm (3ft 3in)
- Tyres: Celerflex 140 x 620; Superflex Artiglio 7.00 x 18; Ultraflex Sigillo Verde 210 x 18
- Engine: SPA CLF, four-cylinder in-line, water-cooled, 1,628cc, 25 HP @2,400 rpm
- Fuel: Gasoline
- Transmission: five speeds forward, one reverse, with reduction gear
- Fuel capacity: 55 litres (14.5 US gallons, 12 Imperial gallons)
- Drive layout: 4x2
- Maximum speed: 38km/h (24mph)
- Range (on road): 300km (186 miles)

Autocarro Sahariano SPA AS 37

An AS 37 from the first series bearing Superflex Sigillo Verde tyres with the classic diamond pattern tread and disc wheels with lightening holes. (Fiat, Viberti)

The interior of the cargo bed. (Fiat, Viberti)

Developmental and Service History

The SPA AS 37 light truck, conceived in 1937, was derived from the TL 37 light tractor and was designed specifically for operations in North Africa. The impetus for its development is attributed to Maresciallo di Campo Italo Balbo, who was Governor-General of Libya, around 1937. The letters AS stood for *Autocarro Sahariano* (Saharan truck), and the vehicle was commonly referred to in Italian simply as the *Sahariano*. As well as the difference in body style compared to the TL 37 tractor, the AS 37 had improved filters, additional fuel and drinkable water tanks and more suitable Superflex Sigillo Verde tyres. A definitive version of the truck appeared in 1939.

Production was somewhat modest, with 1,291 examples totally or partially manufactured before 1942. Production data mixed TL 37s and AS 37s, but 802 trucks were in service on 30 April 1943, with a further 87 awaiting delivery. Following the 8 September 1943 Armistice, the Germans ordered a further 604 vehicles.

A number of SPA AS 37s were employed by the *Compagnie Auto-Sahariana* of the *Battaglione Sahariano*, a specialist unit that combined land and air personnel and vehicles. The *Battaglione* changed its composition between its establishment in 1938 and 1940, on the eve of Italy's entry into the Second World War. In addition to a HQ and a HQ detachment (equipped with light vehicles and machine guns), there was an *Avio-Sahariana Sezione* (Saharan Air Section) with Caproni Ca.309 Ghibli a reconnaissance and ground-attack aircraft; four *Auto-Sahariane Compagnie* (also with light vehicles and machine guns, with liaison, reconnaissance and defence tasks for oases and runways); and a *Meharista Compagnie* (camel cavalry). At the end of

An example equipped with canvas top and Superflex Artiglio tyres, except the spare wheel which is still a Superflex Sigillo Verde. (Fiat, Viberti)

The Autocarro Sahariano AS 37 first series. (Drawing by R. Ciuffoletti and A. M. Feller – GMT)

1942, the military command of the Libyan Sahara converted two AS 37 trucks by eliminating the upper part of their cabs and mounting a 20mm Breda cannon on one example, and a 47/32 gun on the other. During the war, the companies also faced raids by the British Long Range Desert Group.

The AS 37 remained in service with the post-war Italian Army as an artillery tractor for only a brief period, given the wide availability of vehicles of Allied origin.

Technical Description

The AS 37 was laid out conventionally being a 4x4 vehicle with right-hand drive. The cab was all metal and had full doors with windows. The final version had five fuel tanks (one main tank containing 100 litres, two additional tanks with 100 litres each in the cargo bed and another two 50 litre tanks on the roof of the cab) plus four water tanks each containing 50 litres.

The wooden cargo body could carry a full rifle squad of eight men, or a 1,200kg payload and could be covered with a canvas cover; the bed had two seating benches, one each side, which had seats that could be raised to allow access to storage lockers

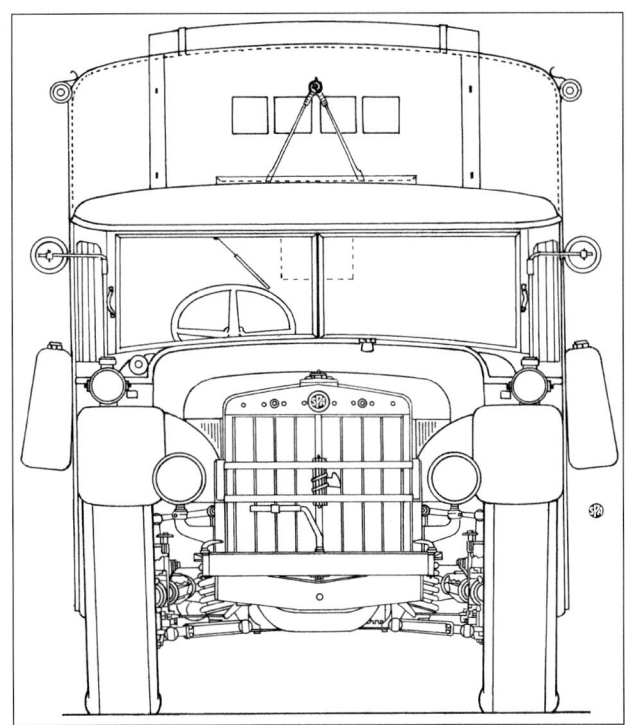
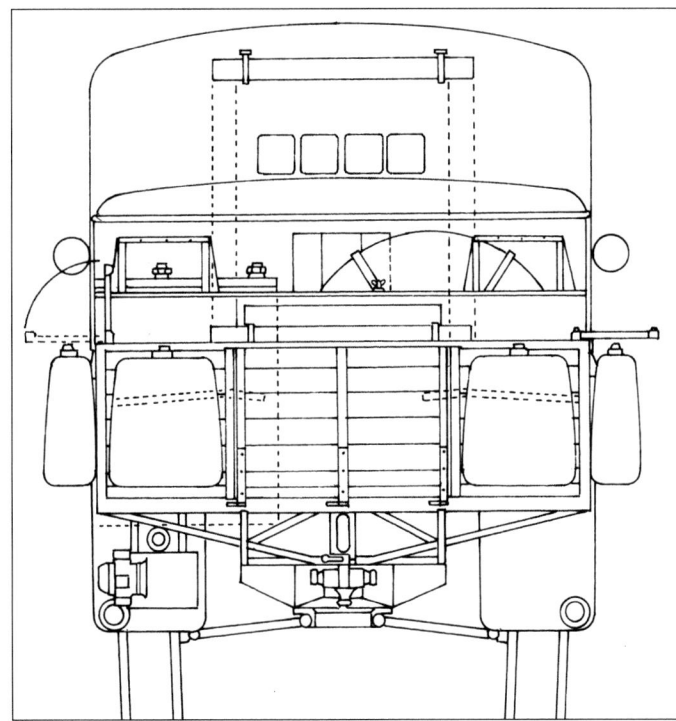

The Autocarro Sahariano AS 37 first series. (Drawing by R. Ciuffoletti and A. M. Feller – GMT)

One of the first examples of AS 37 light truck, at the factory and during trials in North Africa with provisional Superflex Sigillo Raiflex tyres. Note the doors with air vents. (Fiat, Viberti)

An AS 37 from the second series, with standard Sigillo Verde tyres, being tested in the Viberti factory in Turin. (Viberti)

142 ITALIAN SOFT-SKINNED VEHICLES OF THE SECOND WORLD WAR

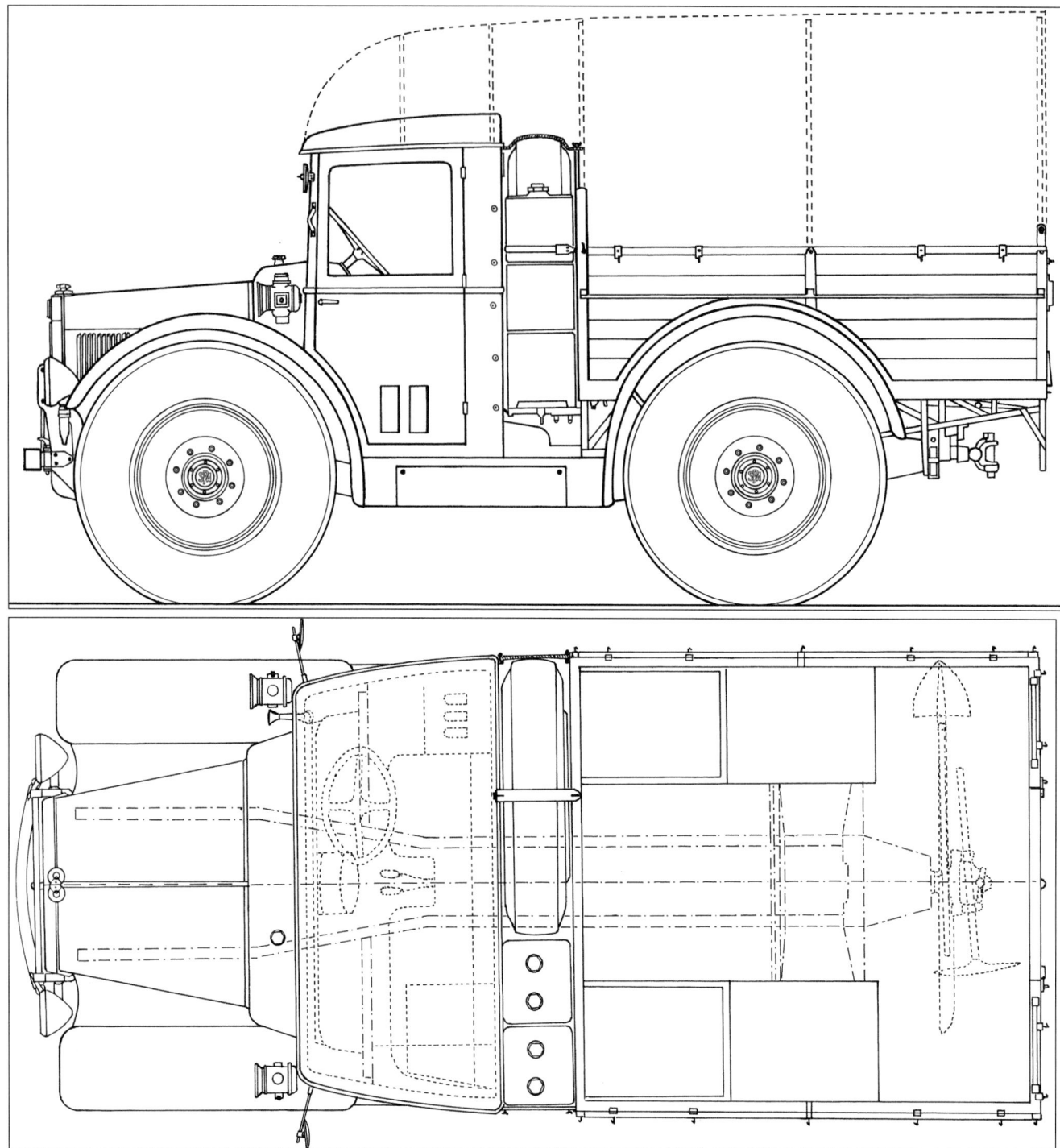

The Autocarro Sahariano AS 37 second series. (Drawing by A. M. Feller – GMT)

beneath. The tailgate consisted of the middle third of the rear panel and, when lowered, had a small integral step to allow easier access to the bed. Early versions of the AS 37 had the same wheels and tyres as the tractor (stamped steel wheels with 8 holes mounting Superflex Artiglio pneumatic tyres), while later production switched to pressed steel wheels with no holes, fitted with Superflex Sigillo Verde or Sigillo Verde Raiflex balloon tyres. A few had pressed steel wheels with Celerflex semi-pneumatic tyres.

The low pressure Superflex tyres enabled it to move easily over desert sand without bogging down. A tyre inflating compressor was carried and a spare tyre was mounted behind the driver's side of the cab. The truck had special oil filters for the engine and transmission. The suspension system was of the type with independent wheels; the front suspension included coil springs with hydraulic shock absorbers, while the rear one consisted of an inverted transversal semi-elliptical leaf spring, more robust compared to the TL 37.

The engine was the four-cylinder Fiat 18T, which was also used on the TL 37 artillery tractor. The manual transmission had four forward and one reverse gears.

A column of AS 37s belonging to a *Compagnia Auto-Sahariana* (Saharan Motorised Company), of the *Battaglione Sahariano* (Saharan Battalion), whose winged lion emblem is visible on the doors. (ACS)

Column of a *Compagnia Auto-Sahariana*.

One of the two AS 37 trucks modified by the *Comando del Sahara Libico* (Libyan Sahara Command or Libyan Sahara HQ) workshop; this example mounted a 47/32 gun. In the background, an AS42 Sahariana.

The second AS 37 modified by the *Comando del Sahara Libico* (Libyan Sahara Command or Libyan Sahara HQ) workshop mounting a Breda 20/65 cannon.

An artillery shell explodes near an AS 37 used as an observation platform.

An AS 37 radio van used by a *Compagnia Marconisti Motorizzati* (Motorised Wireless Operators Company). The van and trailer carried a radio station complete with equipment, antenna and personnel. (ACS)

A radio van with the antenna system deployed. (ACS)

Variants

Variants included a recovery vehicle, a bowser, a mobile repair shop, a closed body radio van and a field version mounting a ladder that could be raised on the rear of the bed and which was used as an artillery observation platform.

A derivative of the AS 37 was the Camionetta Desertica 43 derived from the field experiences in Libya with two modified AS 37 trucks; their closed cab, side doors and windshield were removed and the cargo body of one example was fitted with the 20mm Breda mod. 35 cannon, the second vehicle with the 47/32 gun.

Specifications

- Designation: Autocarro Sahariano SPA AS 37
- Producer: Società Piemontese Automobili (SPA), Turin
- Years produced: 1939–1944
- Number produced: Approximately 1,300
- Length: 4,670mm (15ft 4in)

- Width: 2,020mm (6ft 7in)
- Height: 2,650mm (8ft 8in)
- Unladen weight: 3,770kg (8,311lbs)
- Carrying capacity: 1,200kg (2,646lbs)
- Wheelbase: 2,500mm (8ft 2in)
- Front track: 1,518mm (5ft) with Superflex Artiglio tyres; 1,574mm (5ft 2in) with Superflex Sigillo Verde tyres
- Rear track: 1,518mm (5ft) with Superflex Artiglio tyres; 1,574mm (5ft 2in) with Superflex Sigillo Verde tyres
- Minimum turning radius: 5,000mm (16ft 5in)
- Minimum clearance: 390mm (1ft 3in)
- Fording depth: 700mm (2ft 3in)
- Bed, length: 2,000mm (6ft 7in)
- Bed, width: 1,900mm (6ft 3in)
- Bed, height: 600mm (1ft 11½in)
- Tyres: Celerflex 160 x 88; Superflex Artiglio 9.75 x 24; Superflex Sigillo Verde 11.25 x 24
- Engine: Fiat 18 TL, four-cylinder, water-cooled, 4,053cc, 52 HP @2,000 rpm
- Fuel: Gasoline
- Transmission: four speeds forward, one reverse
- Fuel capacity: 100 litres (26.4 US gallons, 22 Imperial gallons) plus 300 litres (79.25 US gallons, 66 Imperial gallons) in four supplemental tanks
- Drive layout: 4x4
- Speed: 50km/h (31mph)
- Range (on road): 870km (541 miles) with supplemental fuel tanks or canisters

This unidentified vehicle could be the prototype of a repair shop, based on the AS 37 second series but never entered production.

Italian Soft-Skinned Vehicles of the Second World War
Motorcycles, Cars, Trucks, Artillery Tractors 1935–1945

Volume 1

Ralph Riccio

Mario Pieri

Daniele Guglielmi

Helion & Company

Helion & Company Limited
Unit 8 Amherst Business Centre
Budbrooke Road
Warwick
CV34 5WE
England
Tel. 01926 499 619
Email: info@helion.co.uk
Website: www.helion.co.uk
Twitter: @helionbooks
Visit our blog at blog.helion.co.uk

Published by Helion & Company 2023
Designed and typeset by Farr out Publications, Wokingham, Berkshire
Cover designed by Paul Hewitt, Battlefield Design (www.battlefield-design.co.uk)

Text © Ralph Riccio, Mario Pieri, Daniele Guglielmi 2023
Photographs © see Acknowledgements
Colour artwork by and © David Bocquelet 2023

Every reasonable effort has been made to trace copyright holders and to obtain their permission for the use of copyright material. The author and publisher apologise for any errors or omissions in this work, and would be grateful if notified of any corrections that should be incorporated in future reprints or editions of this book.

ISBN 978-1-804513-27-9

British Library Cataloguing-in-Publication Data.
A catalogue record for this book is available from the British Library.

All rights reserved. No part of this publication may be reproduced, stored in a retrieval system, or transmitted, in any form, or by any means, electronic, mechanical, photocopying, recording or otherwise, without the express written consent of Helion & Company Limited.

For details of other military history titles published by Helion & Company Limited contact the above address, or visit our website: http://www.helion.co.uk.

We always welcome receiving book proposals from prospective authors

Contents

List of colour plates	iv
Acknowledgements	v
Documentation sources	vi
Authors' Notes	vii
Abbreviations	viii
Glossary	ix
Foreword	xi
Prefixes and suffixes used in vehicle designations	xii
Introduction	13
1 Overview and Explanatory Notes	17
2 Motorcycles	30
3 Motor Cars	55
4 Light trucks	98

A bibliography will appear in Volume 2

List of colour plates

Fiat 508 CM (aka Fiat 1100 Mimetica) staff car, Italy, 1939. (Artwork by and © David Bocquelet)	I
Autocarretta 35 light truck, mountain chain of the Alps, Italy, 1935. (Artwork by and © David Bocquelet)	I
Fiat 618 light truck, East Africa, 1936. (Artwork by and © David Bocquelet)	II
SPA 38R light truck, Eastern Front, 1941. (Artwork by and © David Bocquelet)	II
SPA CL 39 light truck, Eastern Front, 1941. (Artwork by and © David Bocquelet)	III
SPA AS 37 (second series) light truck, North Africa, 1942. (Artwork by and © David Bocquelet)	III
Benito Mussolini and Marshal Pietro Badoglio on board a Bianchi VM6 C staff car in a newsreel frame from 28 June 1940.	IV
Photograph taken by a German soldier during the transfer trip and stay in North Africa. On the left, an Opel Olympia 38 that appears camouflaged in the *Tropen* colours employed by the Wehrmacht since March 1942. On the right, a Fiat 508 C Camioncino belonging to the *Regio Esercito* whose grey-green colour is only partially covered by the *kaki sahariano* (sand-yellow) applied by spray gun.	IV
A 508 L Camioncino, painted grey-green.	V
An advertising poster for the Breda company (1931).	V
The cover of a 1934 motorcycling magazine with Sertum company propaganda.	VI
The original brand of coachbuilder Viberti.	VI
The updated version of the Viberti brand.	VI

Acknowledgements

The authors and the publisher would like to give thanks to the following individuals for the precious collaboration granted in the creation of this book: Thomas Anderson, Massimo Bartolini, Luigi Carretta, Flavio Chistè, Stefano Di Giusto, Enrico Finazzer, Andrea Olivero, Claudio Pergher, Alberto Pirella, David Zambon, Ennio Zanetti. In particular, we are grateful to the Gruppo Modellistico Trentino di Studio e Ricerca Storica (www.gmt.tn.it – info@gmt.tn.it) and Claudio Pergher again for providing photographs, drawings and other material from its publications and to Flavio Chistè for his constant assistance.

Documentation sources

The documents that appear in this volume come from the repositories and archives cited in the text, or from the authors' private archives. The authors are willing to settle any form of copyright issues pertaining to images and documents, the original provenance of which it has been impossible to determine.

ACS	Archivo Centrale dello Stato, Italy
AUSSME	Archivio Ufficio Storico Stato Maggiore Esercito, Italy
BAMA	Bundesarchiv, Militärarchiv, Germany
CSM	Centro Studi della Motorizzazione, Italy
CTM	Centro Tecnico della Motorizzazione, Italy
ECPAD	Établissement de Communication et de la Production Audiovisuelle de la Défense, France
LIFE	*Life* magazine, USA
GMT	Gruppo Modellistico Trentino di Studio e Ricerca Storica, Italy
ISR	Istituti Storici della Resistenza, Italy
IWM	Imperial War Museum, UK
MMM	Museo della Motorizzazione Militare, Italy
MSGR	Museo Storico Italiano della Guerra di Rovereto, Italy
NARA	National Archives and Records Service, USA
USMM	Ufficio Storico della Marina Militare, Italy

Authors' Notes

The importance of land transport vehicles within an armed force is often underestimated by the average reader, attracted by more 'martial' subjects such as tanks and artillery. Nevertheless, it was thanks to motorcycles, cars, trucks and tractors that – since the early years of the twentieth century – men, weapons, ammunition, provisions, fuel, equipment and orders were transported, all elements without which AFVs, guns and infantry are unable to fight.

In this book we focus on the means of transport in force in the Italian Royal Army (and, in some cases, also in the Italian Royal Air Force and Navy) from the 1930s to the end of the Second World War. Little has been said about them in recent years, even in Italy, with some exceptions such as the Guzzi Alce motorcycle, the Fiat 508 CM car, the Fiat 626/666 and Lancia 3Ro trucks, and a few light and medium tractors.

It is common opinion that the Italian Army was beaten above all because of the poor quality of its combat vehicles. Actually, impartial and in-depth studies made since shortly after the end of the war, have revealed that the main problem was the shortage of vehicles, as well as an entirely insufficient logistics chain. The tank crews were able to compensate with bravery and experience for the fact that their tanks were, from a certain point on, inferior to those of their enemies, but the inadequate number of AFVs and other material was impossible to remedy. The same problem plagued the entire sector of military soft-skinned vehicles, a sign of Italy's limited industrial capacity (and rather of procurement of raw materials and components) compared, for example, to its ally Germany. There were too few factories, too few skilled workers and poor management skills within the armed forces.

However, if quantity was lacking, the same cannot be said for quality. Many models of efficient, robust and resistant vehicles were produced, especially in the sector of the so-called 'standardised' motor vehicles, such as those mentioned at the beginning and others that we will see. These vehicles allowed Italian troops to move and fight in the large and difficult territories of North Africa, the Balkans and the Soviet Union and which brought home what was left of the defeated soldiers.

Abbreviations

AOI	Africa Orientale Italiana (Italian colonies in East Africa)
AS	Africa Settentrionale (North Africa)
MdS	Milizia della Strada, Milizia Nazionale della Strada (National Road Militia)
MVSN	Milizia Volontaria per la Sicurezza Nazionale (Volunteer Militia for National Security)
PAI	Polizia dell'Africa Italiana (Italian African Police)
RA	Regia Aeronautica (Royal Italian Air Force)
RCTC	Regio Corpo Truppe Coloniali (Royal Corps of Colonial Troops)
RCTL	Regio Corpo Truppe Libiche (Royal Corps of Libyan Troops)
RE	Regio Esercito (Royal Italian Army)
RM	Regia Marina (Royal Italian Navy)

Glossary

Italian terms
Aerflex: very low pressure 'balloon' pneumatic tyre for cars and light vehicles
Artiglio: pneumatic tyre with large treads ('artiglio' literally means 'claw')
Autobus, autocorriera: bus
Autocarro: truck, lorry
Autocarro unificato medio: standardised medium truck
Autocarro unificato pesante: standardised heavy truck
Autogruppo: transportation battalion, consisting of several *autoreparti* (transportation companies)
Autoraggruppamento: echelon consisting of several *autogruppi* (transportation battalions); transportation brigade
Autoreparto: transportation company, consisting of several *autosezioni* (transport sections)
Autosezione: motor transport section
Autovettura: motor car
Autovettureta: compact motor car
Biposto: two-seat motorcycle
Camion: truck, lorry
Camioncino: small truck
Camionetta: light truck
Carrello elastico: a single-axle two-wheel bogie trailer on which antiquated artillery pieces were placed in order to make them suitable for high-speed towing
Carro: initially the term indicated a truck or lorry; later it was used for the *carro armato* (tank) and therefore the truck became *autocarro* to avoid misunderstandings
Celerflex: a type of semi-pneumatic tyre
Cingolato: tracked
Cingoletta: light tracked vehicle
Coloniale, Col.: modified versions of a civilian or military vehicle in order to operate in desert and tropical environments, typical of Italian African colonies
Cord: standard pneumatic tyre for light and heavy vehicles
Furgoncino: small van
Furgone: van
Grigioverde: standard grey-green colour for Italian Army materials in the European theatre
Kaki sahariano (or *giallo sabbia*): standard sand-yellow colour for Italian Army equipment in the North African theatre from middle of 1941
Leggero: light
Medio: medium
Militare, Mil.: military
Milizia Marittima di Artiglieria (MILMART): naval artillery militia, normally coastal artillery, but also manned truck-mounted naval guns
Milizia Nazionale della Strada (MdS): branch of the *Milizia Volontaria per la Sicurezza Nazionale* with traffic police functions
Milizia Volontaria per la Sicurezza Nazionale (MVSN): Volunteer Militia for National Security, a Fascist militia organisation
Modello, Mod.: model
Monoposto: single-seat motorcycle

Motocarrozzetta: motorcycle/sidecar combination
Motocicletta, motociclo: motorcycle
Mototriciclo: three-wheeled motorcycle
Pesante: heavy
Polizia Coloniale: original name of the *Polizia dell'Africa Italiana*
Polizia dell'Africa Italiana (PAI): Italian African Police, the police corps of Italian North Africa and Italian East Africa colonies from 1936 to 1945, reporting directly to the Ministry of the Colonies, later renamed the Ministry of Italian Africa. Until 1939 the corps was named *Polizia Coloniale*
Regia Aeronautica (RA): Italian Royal Air Force
Regio Corpo Truppe Coloniali (RCTC): Royal Corps of Colonial Troops, Italian military unit which included troops stationed in the African colonies of Eritrea, Somalia, Tripolitania and Cirenaica (Tripolitania and Cirenaica were later merged into Libya)
Regio Corpo Truppe Libiche (RCTL): Royal Corps of Libyan Troops (1938–1943); formerly Regio Corpo Truppe Coloniali della Libia (1935–1938)
Regio Esercito (RE): Italian Royal Army
Regia Marina (RM): Italian Royal Navy
Rimorchio: trailer
Semicingolato: half-track
*Superflex (*aka *Superflex Cord):* low pressure pneumatic tyre for light and heavy vehicles
Sigillo Verde: special tread for heavy vehicles tyres suitable for soft soils
Sigillo Verde Libia (aka *Tipo Libia*): special tread for tyres designed for sand
Stella Bianca: special tread for light vehicles, similar to the *sigillo verde*
Trattore: tractor
Trattrice: tractor (usually heavy)
Ultraflex: very low pressure 'balloon' pneumatic tyre for heavy vehicles

Foreword

I am grateful to write this section for the book, *Italian Soft-Skinned Vehicles of the Second World War: Motorcycles, Cars, Trucks, Artillery Tractors 1940–1945* by Ralph Riccio, Daniele Guglielmi and Mario Pieri. I feel as such, not only because it is an honour to be asked by these well-respected authors, but also because this book is needed. There is very little English language reference material covering these subjects which is available to military vehicle enthusiasts and model hobbyists.

Since first becoming interested in military vehicles as a boy, my primary interest for the past twenty years has become armoured cars and wheeled fighting vehicles. Although related, I must admit I was not terribly predisposed towards soft-skinned vehicles. However, over the years as I became more knowledgeable about armoured cars, my interest could not but be drawn towards the types of vehicles covered in this book. The obvious automotive background and technical similarities shared by both armoured cars and soft-skinned vehicles almost made it a necessity that I focus more on the latter, to better understand the former.

During my experiences, needless to say, I have found that there is a vast amount of printed and online references available covering armoured fighting vehicles. Although, I have not been terribly disappointed in my search for information covering soft-skinned vehicles, it is harder to come by than with other some other vehicles. In addition, as one will recognise quickly while learning or researching military history in general, and equipment and vehicles specifically, some countries will often be the subject of more works than others. Unfortunately, Italy seems to be one of the nations covered less than others, especially by English language sources. This situation is even more skewed when it comes to Italian military vehicles, specifically soft-skins. As a result, I have found it difficult to locate high-quality English language reference material covering this subject.

So needless to say, as a military enthusiast and model hobbyist, I am very pleased that this book has been written by these noted authors and published for our use. It is a welcome addition to my reference collection. I also can assure you, it won't be collecting dust on my shelf, as I will be referencing it often.

Patrick Keenan
Editor – WarWheels.net
September 2023

Prefixes and suffixes used in vehicle designations

AS	Autocarro sahariano (desert truck)
B	Benzina (petrol, gasoline)
BM	Benzina, Militare (military vehicle with petrol engine)
C	Coloniale (tropicalised vehicle or equipment)
C	Corto (short wheelbase)
CL	Carro leggero (light truck)
CM	Carro militare (military truck)
CV	Carro veloce (fast tank, actually a tankette)
G	Gassogeno (gas generator engine)
GM	Gassogeno, Militare (military vehicle with gas generator engine)
L	Lungo (long wheelbase)
M	Militare (military vehicle)
N	Nafta (diesel oil)
NM	Nafta, Militare (military vehicle with diesel engine)
P	Pneumatici (pneumatic tyres)
PC	Pesante Campale (heavy field)
R	Ribassato (lowered chassis for use for buses and other special vehicles)
S or SP	Semipneumatici (semi-pneumatic tyres)
TL	Trattore leggero (light tractor)
TM	Trattore medio (medium tractor)
TP	Trattore pesante (heavy tractor)

Introduction

Part (perhaps a very large part) of the reason that the *Regio Esercito* did not perform as well as it might have otherwise have done during the Second World War was that it lacked sufficient mobility, compromising its ability to perform competitively on the battlefield against opponents that were more highly motorised and mechanised. Limited mobility was a direct result of the fact that the Italian automotive sector was never able to provide the number of trucks required to support the armed forces, especially the *Regio Esercito*. Despite the fact that Italian infantry divisions were much smaller than those of the other major combatant nations (the standard infantry division was a 'binary' division consisting of only two, rather than three, infantry regiments), most Italian infantry divisions were chronically short of trucks, especially enough trucks to move the troops, and barely had enough trucks and tractors to equip the artillery regiments within the division. As early as the opening moves of the desert war between Italy and Britain, prior to September 1940, Marshal Rodolfo Graziani, who was acutely aware of the deficit of motor transport in North Africa, pleaded with Mussolini for 600 additional trucks so that he could completely motorise his attack force. Mussolini denied the request and ordered that the planned attack proceed without the trucks. The situation did not improve as the war progressed: in July 1941, there were only about 5,200 trucks available to an Italian force in North Africa consisting of 110,000 men.

The reasons for this woeful shortage of vehicles were many and varied; foremost among these reasons was that Italy's industrial base was simply too small to be able to adequately provide all of the vehicles required to equip an army consonant with the times. One of the endemic problems with the Italian industrial base that produced equipment for the military forces was that the too few companies were spread too thin, producing hardware for all three services; as examples, Fiat built aircraft, tanks and motor vehicles while Ansaldo built ships and artillery among other items. Other factors that affected production

An Italian driving licence assigned on 26 January 1916 to a Corporal of the *87° Reggimento Fanteria* of the *Regio Esercito*.

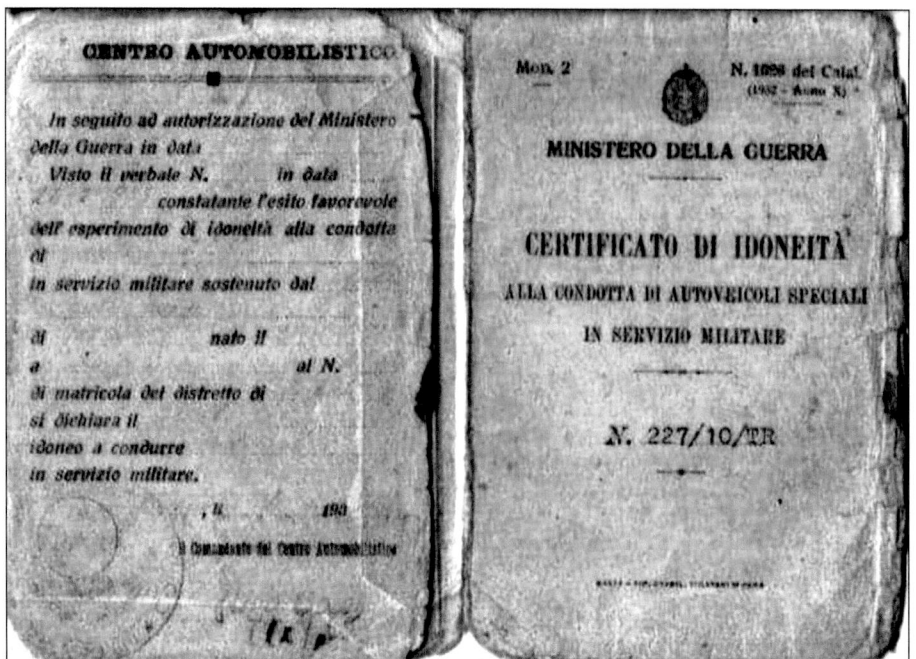

A blank driving licence issued by the War Office in 1932.

The heterogeneity of the fleet of logistic vehicles used by the Italian Army is exemplified by these trucks of an Autosezione marching in Libya in the winter of 1940–1941. From left: an old Fiat 618, a Fiat 634 N2 and a SPA Dovunque 35. (ACS)

Fiat 500 Topolino under construction in the Fiat Mirafiori plant, which had been opened in 1939. (Fiat)

were a scarcity of primary materials to support the entire war effort and the lack of an orchestrated effort to coordinate and maximise existing production capabilities.

Among the automotive companies suitable for the Italian military, the high-volume producers (in a relative sense) were Fiat, and its subsidiaries SPA and Ceirano. Compared with this 'giant consortium', Lancia, Breda, Isotta Fraschini, Officine Meccaniche (OM) and Bianchi were somewhat lower-volume car and truck manufacturers. In addition to producing cars and trucks (both civilian and military), the Fiat group also produced tanks, engines and aircraft. Likewise, Alfa Romeo built aero engines, Isotta Fraschini built aero engines and marine engines for motor torpedo boats as well as anti-aircraft guns, and Breda produced railway equipment, and light weapons and anti-aircraft guns. Almost inevitably, and not surprisingly, when needs were prioritised, soft-skinned motor vehicles came in towards the bottom of the pile. And as if that wasn't enough, these companies were in competition and rarely worked together. The General Staff of the Italian Army calculated in 1938 that the aforementioned companies could monthly produce 650 heavy trucks, 600 light trucks, 300 vehicles of various types (mainly tractors), 100 *autocarrette* and 50 heavy tractors, for a total of 1,700 motor vehicles, to be reduced to 1,500 considering the needs of the other Italian armed forces. Actually, these figures were achieved only for short periods.

The result was that, although tanks and guns may have taken priority over trucks, without trucks (and tractors) the artillery and its ammunition could not be moved, the tanks could not be transported to where they were needed, and the infantry would have to sit where it was and could not be moved in time

A column of motor vehicles of the Italian *Regio Esercito* moving towards the front lines of the Eastern Front through a town in the summer of 1941. (ACS)

to influence the course of a battle. Given the scarce resources at hand, the Italian military and industry faced an impossible dilemma when it came to trucks or combat equipment. In oversimplified terms, if Italy produced a greater of number of trucks at the expense of combat equipment, there would have been little or no equipment such as artillery to transport, whereas if the priority had been given to major combat items (as was the case), there would not be enough trucks, or tractors, to adequately support the combat arms.

Perhaps one bright spot in the Italian vehicle production picture was the number of motorcycles the various Italian manufacturers were able to produce to meet military requirements. Motorcycles required relatively few raw materials and did not consume much fuel per vehicle, but the downside was that a motorcycle could carry only one or two soldiers, which meant that it took approximately ten to twenty motorcycles to carry as many men as a single truck, somewhat negating the lower cost and fuel savings per motorcycle.

Although an imprecise yardstick, it is illustrative to consider a few figures regarding Italian truck production compared to production in a similar timeframe in Germany and the United States. During the Second World War, Italy had an army of roughly 4 million men (ground forces only, excluding navy and air forces), while Germany had around 13 million and the USA around 11 million, so both Germany and the USA had approximately three times as many soldiers as Italy. As the following illustrates, even accounting for the relative sizes of the armies involved, there was nevertheless a huge disparity between Italian truck production and that of Germany and the USA.

Before the outbreak of the Second World War, according to official sources the Italian Army had about 22,000 motor vehicles in service (6,487 light trucks, 5,887 heavy trucks, 2,332 *autocarrette*, 2,921 special vehicles, 4,328 tractors). By June 1940 the total had risen to 53,000 vehicles, excluding tanks, of which 17,000 were requisitioned in previous years. In June 1943 the armed forces (Army, Navy and Air Force) had 96,000 vehicles and 40,000 motorcycles, most of them manufactured in Italy.

Nicola Pignato (see bibliography) estimated that between 1940 and 1945, Italy produced about 163,200 motor vehicles (cars, trucks and tractors). By comparison, between 1935 and 1944, Germany produced

Type of truck	Load		Speed in kilometers per hour	
	Men	Material (kg)	Single vehicles	Columns (average)
Light:				
Infantry Truck	10 to 12	1,000	40	25
OM4 OMF	12 to 15	1,200	60	30
Fiat 618 CM	12 to 15	1,250	65	30
Spa 25 C 10	16 to 20	1,800	50	25
Fiat 612 P	20 to 25	2,500	43	25
Spa 38 R	20 to 25	2,500	52	25
Ceirano 47 CM	20 to 25	3,000	45	25
Bianchi Mediolanum 36	20 to 25	3,000	55	25
OM, CRD	20 to 25	3,000	51	25
Isotta Fraschini D 70 NM	20 to 25	3,000	56	25
Heavy:				
Ceirano 50 CM	20 to 25	5,000	25	18
Lancia RO.NM	24 to 30	5,000	32	20
RO BM	24 to 30	5,000	39	22
Fiat 633 NM	24 to 30	5,000	30	20
Fiat 633 GM	24 to 30	3,500	28	18
Isotta Fraschini D 80 NM	24 to 30	5,000	34	20
OM 3 BOD	24 to 30	5,000	51	23
Giant:				
Fiat 634 N	28 to 32	7,000	40	20
3 RO Lancia	28 to 32	6,500	43	22

Table of motor transport loads, from *TME 30-420 Handbook on the Italian Military Forces* (3 August 1943).

approximately 130,000 Opel Blitz trucks alone, while the figures for US truck production are simply staggering in comparison. US automotive manufacturers churned out more than 200,000 Studebaker US 6 trucks (the US 6 was actually manufactured by REO as well as by Studebaker, both relatively small companies by American standards) most of which were provided to the Soviet Union, as well as 572,500 GMC CCKW-series 2 half-ton trucks and 382,350 Dodge WC-series three-quarter-ton trucks, not to mention 647,925 Willys MB and Ford GPW jeeps – almost two million vehicles, excluding special vehicles and heavy 6x6 trucks built by manufacturers such as Diamond T, White, International Harvester, Autocar, Mack, Brockway, Corbitt, FWD and Ward La France which numbered well into the hundreds of thousands.

In the years leading up to the Second World War, the British government prepared the industry for the mass production of required motor vehicles. Britain entered the war with 80,000 military vehicles of all types, although most were left behind in the evacuation at Dunkirk in 1940. The Canadian auto industry not only replaced these losses, it did much more, producing more than 800,000 military transport vehicles.

Unlike its German ally which plundered all of the territories it occupied for anything related to the war effort, including requisitioning of motor vehicles as well as of the factories that produced them (Renault, Citroën and others in France; Tatra and Skoda in Czechoslovakia), the countries that Italy was able to occupy during the early stages of the war (Albania, Greece, the Balkans, East Africa and North Africa) yielded no such windfalls. Requisitioning civilian vehicles in the Italian colonies or Italy itself did little to fill the needs of the military; in 1940, a plan by the army to requisition 20,500 civilian trucks fell short by 7,900. It is useful to note that the availability of vehicles in Italy prior to the war was quite limited compared to the vehicle numbers registered in nearby countries; in 1939, Italy had only about 300,000 cars, or about one for every 130 people, whereas neighbouring France had almost 2 million cars, or one for every 23 people.

Historians have discussed a lot about the reasons why a direct and fruitful collaboration between the German and Italian factories never materialised, despite the alliance between the two nations and the proclamations of the two leaders and their main collaborators. Basically, as Lucio Ceva and Andrea Curami have described (see bibliography), the main shortcoming was on the Italian side. Professional jealousies, fear of losing commercial influence, political and industrial manoeuvres meant that, for example, no German engine of adequate power to be able to move tanks and heavy vehicles arrived in Italy, let alone finished products. At the same time, only a few vehicles requisitioned by the Germans in the occupied territories were forwarded to the *Regio Esercito*.

It is also necessary to underline the Italian war effort in the wars of the thirties in East Africa and (even if not officially) in Spain. These commitments drained the nation's military funds and reduced the circulating fleet of transport vehicles, as the army resorted to purchased or requisitioned civilian vehicles for reasons of time and resources.

The dire shortage of trucks in the Second World War was further compounded by their performance characteristics, which by comparison fell below those of some of the US trucks mentioned above. The GMC CCKW was a 6x6 truck with a 91.5 HP engine, a range of 300 miles (482km), and a road speed of 45mph (72km/h), while Italian trucks such as the Fiat 626 and Bianchi Miles had engines of about 65 HP, had a range of 400km (249 miles) and a road speed of about 63km/h (40mph). In fact, as mentioned above, the low power and, at times, reliability of some Italian engines, especially diesel engines, was a very significant 'Achilles' heel'. This in no way detracts from the overall quality of Italian trucks in general, but the lower performance characteristics vis-à-vis trucks of some other nations ultimately translated to even less ability to move troops and equipment as expeditiously as needed.

1

Overview and Explanatory Notes

Historical context

It can be said that, during the 1930s, the interest of the Italian Army was concentrated on two aspects: the defence of the country's land territorial borders, made up entirely of the mountain chain of the Alps, and the conquest and/ or administration of colonial territories in North and East Africa.

In the first case, it was evident that in the event of a war the main effort would have been performed by infantry, mountain troops and artillery, so the military authorities decided to adopt a light and very mobile tank – essentially a tankette – such as the Carro Veloce CV 33 (later the CV 35 and CV 38 versions), to be used as a fire support system. The army also equipped itself with light trucks (called *autocarrette*) with a maximum speed of only a few km/h but theoretically capable of traversing the mule tracks in the mountains, and wheeled artillery tractors suitable for towing field and siege guns. In the colonies, however, armoured cars, including those

An *autocarretta* (light truck) while facing a steep slope during the summer manoeuvres of 1932.

A column of military motor cars, headed by a Lancia Aprilia Coloniale, together with a sidecar and other Italian vehicles in a North African town. (D. Zambon)

Fragment of an advertising poster from 1939. The caption explains: 'FIAT 666 N. The new heavy-duty truck. Standardised type. It will surpass the success of its famous predecessor: the 634 N'.

that had survived from the Great War, were more useful, together with motor vehicles for the transport of men and materials.

With regard to the latter category, procurement was based on what was offered by domestic manufacturers. Only towards the end of the decade did the so-called *coloniale* (colonial) versions of many cars and trucks appear; these were modified versions based on standard models, intended both for the civilian and military market in order to operate in the desert and tropical environments typical of Italian African colonies.

At the time, Italian industry was still able to offer models of good overall quality, sturdy and quite reliable; at least as regards their use on the national territory, rich in hills and mountains. In spite of this, the army had mostly old and very diverse vehicles in active service, such as the Fiat 18 BLR and Fiat 15 Ter trucks. Italy's victories in the Italo-Ethiopian War (1935–1937) and in the Spanish Civil War (1937–1939) hid the existing problems relating to the numbers of vehicles available. In 1939 there were fewer than 50,000 examples serving in the army: old, new, civilian and specifically military. In fact, when Italy's entry into the war alongside Germany in June 1940 required a greater number of motor vehicles – not just AFVs – Italian industry revealed its limits. The shortage of qualified personnel (also because many men had been called up to the armed forces), the difficulty in finding materials and components, the scarcity of public funds because much had been spent in previous wars, the rivalry between companies, and finally political intrigues were among the main causes behind a production that was much lower than needed.

Campaigns of the Italian armed forces in North Africa and the Soviet Union, still alongside their German ally, in addition to operations in the Balkans, often put the entire logistics chain in crisis. Thanks to the self-denial and sacrifice of many crews and soldiers, the army still managed to fight.

After the Armistice of 8 September 1943, the Germans took over the entire Italian war production, controlling the factories. It is obvious that any type of vehicle was precious at the time; however Mussolini's former ally benefited from the availability of many Italian cars, trucks and tractors as it continued its war against Allied troops.

Autocarri Militari Unificati (Standardised Military Trucks)

During the war in Ethiopia, the Italian armed forces used a considerable variety of motor vehicles, including those of foreign manufacture. The vehicle fleet existing in Italy at that time was varied and, in the event of general mobilisation, it would have been possible to requisition many thousands of cars and trucks of different types and brands. This would perhaps have solved

A service station in 1935. (Fisogni Museum)

The first Fiat diesel engine for trucks, the four-cylinder model 350 with direct injection of 5,570cc mounted on the Fiat 632 N. (Negri Foundation)

the urgent needs, but on the other hand complicated the logistic situation even more.

Therefore, in July 1937 a decree was issued to concentrate production as much as possible, reducing the number of civilian vehicles belonging to the categories of light, medium and heavy trucks. The Italian manufacturers were told what common features these vehicles should have, in terms of engines, transmissions, drive wheels, cabs, performance, fuel capacity, weights; they would have 18 months to adjust production, otherwise the vehicles would not be admitted into circulation. In November of the same year, with the publication of the official rules, the concept of the *automezzo unificato* (standardised) motor vehicle was thus born.

The new classification of vehicles was based mainly on the engine type, the total weight and the maximum load. Newly built 2 axle trucks had to fall into one of the following three types:

1. heavy trucks with diesel engine, maximum laden weight of 12,000kg and maximum load of at least 6,000kg;
2. medium trucks with a maximum laden weight of 6,500kg and maximum load of at least 3,000kg;
3. light trucks with a maximum laden weight not exceeding 4,000kg and maximum load of at least 1,500kg.

This categorisation received some changes over time.

Even the main mechanical parts – from bolts to leaf spring leaves for the suspensions – were to have been interchangeable. The standardisation also covered towing pintles, braking and electrical systems, tyres and headlights. Also recommended was the adoption of diesel engines instead of gasoline engines, with low revs and high torque values. The placement of door windows in canvas and mica or celluloid instead of glass had to be provided for. It is possible that the decision, although logical in itself, was inspired by the analogous programme of Italy's German ally who, a short time before, had imposed rules for the production of *Einheits* type (standardised pattern) vehicles, with simplified parts.

Due to the participation of the Italian Expeditionary Forces in the Spanish Civil War, there was neither time nor opportunity for manufacturers to put the recommendations into practice, and this further revealed the complications that existed in managing such a heterogeneous fleet. As in other cases, in fact Italian industry adapted very slowly to the directives of the military authorities, fearing that much of its production could be requisitioned rather than sold. This situation persisted until the outbreak of the Second World War, so much so that when Italy entered the war on 10 June 1940, standardisation was still a long way off.

In any case, some progress had been made: the models in production were fewer than a few years before and had similar characteristics. However, it should be noted that, despite the use of old motor vehicles – even veterans of the Great War – or requisitioned civilian vehicles, as well as those purchased abroad or war booty once hostilities began, the needs of the Italian armed forces were never satisfied, leaving an endemic shortage that had serious consequences on the mobility and efficiency of the troops on all fronts. Most of the units were not independent in terms of logistics vehicles and had to use special motorised detachments (named *autoraggruppamenti, autogruppi, autoreparti* and *autosezioni*, at different levels) at the disposal of the armies and army corps; but they never had a sufficient amount of equipment to transport men and materials as required for military operations. Many trucks and their drivers worked tirelessly, until a mechanical failure, an accident or enemy action stopped them.

From the chassis of the standardised trucks, many derivatives such as buses, vans, tank trucks, ambulances, radio vehicles and mobile workshops could be obtained. Normal traction was on a single drive axle, the rear one, even though the design of four-wheel drive vehicles was encouraged – a feature much requested by the military units.

These standardised vehicles were mainly cars and trucks, but there were also trailers and motorcycles; they were widespread during the Second World War and in the years following. The standardised trucks included in the medium weight category the Fiat 626, perhaps the most famous, together with Bianchi Miles, Alfa Romeo 430, OM Taurus, Isotta Fraschini D65, and Lancia Esaro (this latter only entered production in mid–1943). The heavy category included the Fiat 666, Lancia 3Ro, Isotta Fraschini D80, Alfa Romeo 800, and OM Ursus.

Engines and fuels

Since the end of the nineteenth century, internal combustion engines were widespread in Italy for civil, industrial, agricultural and military uses. It was, after all, two Italians Niccolò (Eugenio) Barsanti, a teacher of mathematics and physics born in Pietrasanta in 1821, and Felice Matteucci, an engineer born in Lucca in 1808, who were the first in the world to design and patent an internal combustion engine. It was 1854 and two years later they built a prototype in Florence that used a mixture of air and illuminating gas (illuminating gas was made from bituminous coal and was widely used for municipal lighting).

Refuelling of Fiat Dovunque and other trucks in Libya at the end of 1940. (ACS)

Refuelling Fiat 634 N2 trucks belonging to the *271ª Autosezione*. (ACS)

In the early twentieth century, gasoline was the most widely used fuel in Italy for transport vehicles, or more precisely the only one. For example, a manual published in 1922 explains that benzene was not widespread, like ethyl and methyl alcohol, even when mixed with the former. On the other hand, diesel engines were bulky and noisy at the time, so their use was reserved for ships, industrial plants and, later, locomotives.

Only at the beginning of the 1930s engines fuelled by *nafta* (the term used for diesel fuel) began to be mounted on heavy civilian and military trucks, once they could be made smaller and more manageable. In summary, the first diesel engines for vehicles had limited power and their starting was complex. On the other hand, their efficiency was higher than gasoline-powered engines; this allowed lower consumption and, consequently, greater range.

After years of study, Fiat installed the first diesel engine for trucks on the 632 N model (civilian) in 1931; the first Fiat military truck running on diesel was the 633 NM introduced in 1935. Other companies, on the other hand, preferred to build engines of foreign manufacturers under licence, such as OM (under licence from Switzerland's Saurer) and Isotta Fraschini (under licence from Germany's MAN).

Military authorities began to prefer diesel engines, even for soft-skinned vehicles, because they are less liable to flammability, an especially important factor in combat vehicles. In this sense, the important parameter is the 'flash point', which for gasoline is about -20°C (-4°F): this implies that at room temperature it spontaneously emits vapours which are visible to the naked eye and are liable to catch fire with a minimum ignition, i.e. a heat source that provides the thermal energy. Diesel fuel, on the other hand, has a flammability temperature of about 55°C (131°F), so under normal environmental conditions it is much less prone to ignite than gasoline: only if it is heated to above 55°C does the emission of flammable vapours occur. Moreover, the mechanical efficiency was higher with a diesel engine, with a lower fuel consumption.

In Italy, until the outbreak of the First World War, there were few motor vehicles in use; refuelling was done by buying gasoline in various places, such as pharmacies and drugstores. The first fuel pumps were installed where the number of

A column of Lancia 3Ro; the leader is a bowser (tanker truck) and has the cab roof painted with the Italian tricolour flag for aerial recognition.

vehicles was greater, such as in garages, bus terminals and in some luxury hotels. After the end of that conflict, civilian motor vehicles in use in Italy numbered just over 30,000, a third of which were trucks. In WW1 the army operated just over 2,000 cars, all of civilian production, and about 500 specialised vehicles (ambulances, tank truck, buses), as well as a few thousand motorcycles used mainly for communications; on the other hand, the number of trucks exceeded 40,000 and of artillery tractors over 900 (without taking into account the losses).

Italian jerry cans equip a SPA AS 42 Sahariana; their resemblance to the German model is evident.

Gradually, in the twenties, on the basis of first American, then French and German experiences, the first gasoline pumps – called metering pumps – were built and approved, to be placed along some roads and motorways. The following decade was characterised by an economic crisis and, consequently, by a drop in consumption, but with the spread of diesel engines for trucks, even the distribution of fuels saw an evolution. Soon, street filling stations dispensed both petrol and diesel fuel.

Beginning from the first wars waged by Italy in the African territories and then in Spain in the second half of the thirties, the problem of how to transport fuels was a concern of the military authorities. However, for years 200 litre drums were the only container used by the army to supply transport, reconnaissance and combat vehicles. Only towards the middle of 1941, during the North African campaign against the British forces, was the jerry can built on the German model (*Einheitskanister*) adopted. It differed from that of Italy's German ally in having

An Italian jerry can built on the model of an early German type. The white painted 'X' shaped notch indicates that it contains potable water. (ACS)

Italian equipment captured by the Germans after the Armistice of 8 September 1943. Behind are a Guzzi Alce motorcycle, a number of jerry cans and some Fiat 626 trucks. (ECPAD)

the letters 'R.E.' (more rarely, with the letters 'RA' and 'RM') and the wording *20 litri* (20 litres) imprinted on the left side. Jerry cans marked with a white stripe or cross indicated that the contents were drinking water; however, often, as veterans tell us, the water was not drinkable because out of necessity containers that had previously been used to transport petrol or diesel were used.

Tanker trucks (bowsers) made their appearance during the war, often requisitioned civilian vehicles, and 200 litre fuel drums also remained in use.

Wheels and Tyres

The military specifications for acceptance required, at least until the mid-thirties, the presence of puncture-resistant tyres, since punctures were frequent because of the conditions of the roads, which were often scattered with nails lost from the metal shoes of draft animals, and – in the combat zone – from shrapnel, splinters and small arms bullets. At the beginning, leaving out the all-metal wheels and other solutions for tractors, solid rubber ring tyres were chosen. The downside was their rigidity, which made the crew's ride uncomfortable and tiring, also causing vibrations and stresses to the mechanical parts, eventually damaging them. Additionally, solid tyres had a tendency to overheat.

A partial solution to the problem was found with semi-pneumatic tyres, which included one or more cavities, called ventilation chambers, containing air at ambient pressure. In this way they were softer and less prone to overheating. The most modern anti-perforation semi-pneumatic model used by the Italian armed forces was the 'Celerflex', produced by Pirelli from 1937 on the basis of the experiences gained in previous years. Another was the 'Cellastico', made in Italy by Società Italiana Industria Gomma under licence from a Dutch manufacturer supplying various European armies.

Over time, the semi-pneumatic tyres gradually gave way to pneumatic tyres. Military authorities sometimes required a vehicle to fit both types. In fact, the negative aspects of semi-pneumatic tyres mentioned above were highlighted in the military campaigns of the 1930s in North and East Africa and in Spain; for this reason, they were abandoned. At the outbreak of the Second World War all the vehicles of the *Regio Esercito* were fitted with pneumatic tyres, although various photographs from the period show older vehicles still equipped with semi-pneumatic tyres, especially in the European theatre of war.

Semi-pneumatic tyres were also placed on the wheels of the

A 1926 advertising poster for Pirelli semi-pneumatic tyres, 'ideal for heavy vehicles'.

The chassis of a TM 40 medium tractor with Celerflex semi-pneumatic tyres.

carriages of artillery pieces. Cycles and motorcycles, on the other hand, were equipped only with pneumatic tyres.

The rubber tyres used by the Italian military vehicles of the time were, with rare exceptions, all manufactured by the Milanese company, Pirelli, which also had a solid civilian market outside Italy. Its catalogues, of course, varied over time, introducing new models with different construction features, dictated by technological advancement or by contingent needs. Pirelli tyres were designed for both civil and military use, since the armed forces also used vehicles not specifically designed for military purposes. Only after the introduction of standardised vehicles and trailers was there some rationalisation from a military point of view.

Two different Pirelli semi-pneumatic tyres.

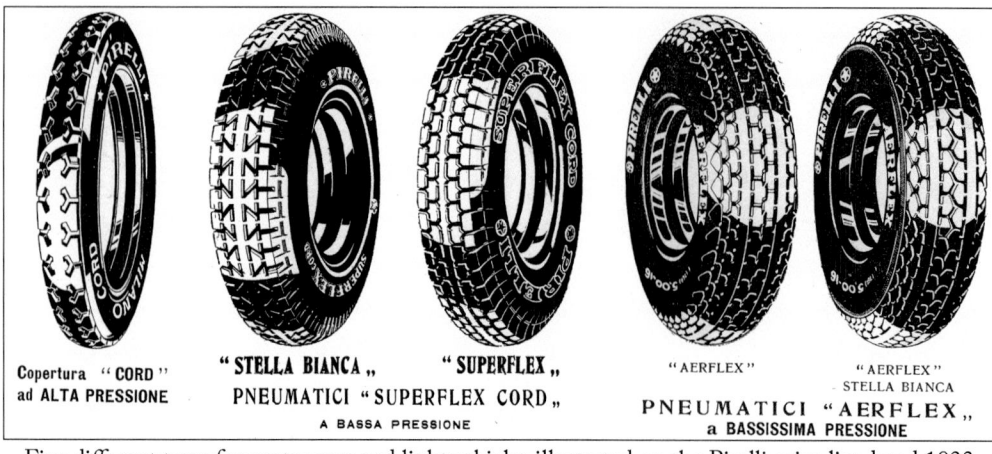

Five different tyres for motor cars and light vehicles illustrated on the Pirelli price list dated 1933. From left: the Cord, also available with artiglio tread (not shown here); the Superflex Cord with increased cross-section and low pressure, in the variants Stella Bianca and normal Superflex treads; the very low pressure Aerflex, also in this case including 'Normal' and 'Stella Bianca' variants.

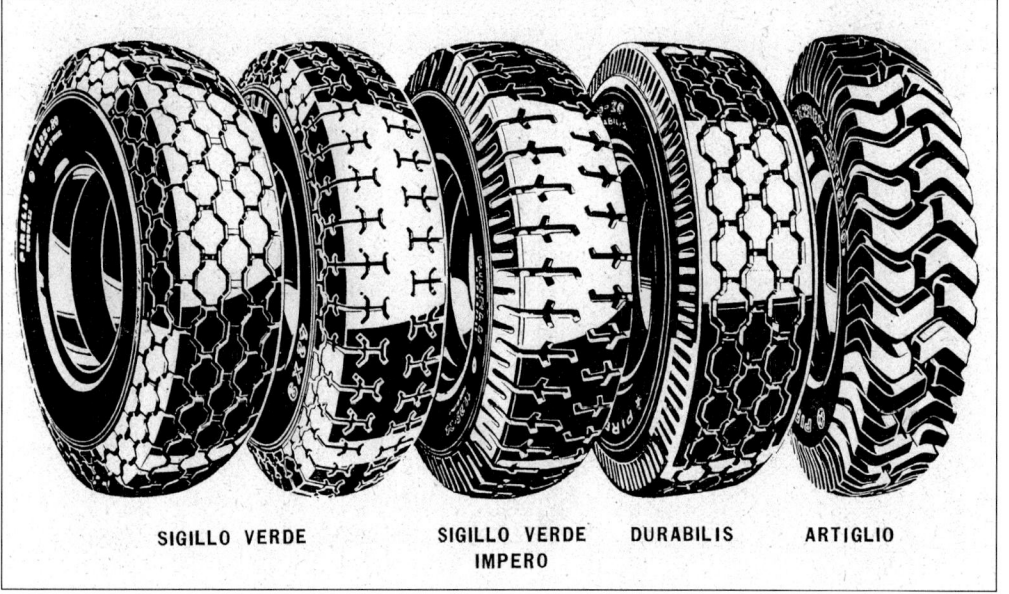

Five different treads for heavy vehicles illustrated on the Pirelli price list dated 1940 and available on the Cord, Superflex and Ultraflex tyre models. From left: the *sigillo verde* in two variants, the first one with the characteristic polygonal mesh engravings, while the second variant is the Raiflex; the Sigillo Verde Impero, featuring the 'f' carvings; the Durabilis for long-distance vehicles; finally the most recent Artiglio tread.

Strictly speaking, we should examine the tyres of all categories of motor vehicles in service in the 1930s and 1940s in the Italian armed forces. However, the list would be long, so we will limit ourselves to mentioning the most popular models and their treads, reiterating that there was no distinction between civil and military models.

In the older models of tyres, the carcass, that is the flexible and inextensible casing that contains the air chamber, was built – in simple terms – by immersing several layers (plies) of knitted cotton fabric in the melted rubber; in this way, the flexibility of the rubber was associated with the resistance to stresses given by the fabric. During the first and second decades of the twentieth century, the cord fabric (short for the French *corduroy*) was introduced; it replaced the common transverse cotton fabric as it is more durable and less subject to heating from friction. The Pirelli model Cord of 1921 represented the type of standard tyre for this manufacturer for many years.

In 1924 Pirelli introduced a new model, the Superflex Cord (or, more simply, Superflex), characterised by a larger section and low pressure inflation, followed later (early thirties) by the Aerflex and the Ultraflex at very low pressure inflation; the Aerflex was intended for passenger cars and light trucks, the Ultraflex for heavy vehicles. A lower pressure corresponded to a lower propensity to punctures, greater comfort and better grip on less compact ground, thanks also to the wider section: tyres of this type were also universally called *balloon*. High-pressure and low pressure Cord and Superflex Cord tyres were also available for two-wheeled and three-wheeled motorcycles.

Towards the end of the decade, these basic models were produced in the Raiflex variant, which had the carcass with a rayon fabric, a synthetic textile fibre derived from cellulose; in this way the Italian industry tried to reduce its dependence on imports, in this case of cotton. In addition, rayon guaranteed greater uniformity and resistance thanks to the structure of its fibre.

An aesthetic and functional differentiation among tyres was given by the tread, that is the part that makes contact with the ground.

One of the most common treads for the aforementioned Pirelli 'flex' models was the *Sigillo Verde* (Green Seal) patented in 1930; the manufacturer offered it in different sizes suitable for both wheels for cars or light vehicles and for heavy transport vehicles (in this case they were called 'giant tyres'). The sigillo verde tread was lightly sculpted, with a pattern that varied depending on the model of tyre that it was on. Apparently the most widespread pattern – at least for military vehicles – had almost hexagonal

A photograph from the *Salone dell'Automobile* (Motor Show), held in Milan in 1939. The Stella Bianca and Sigillo Verde tyres are displayed on the Pirelli stand. (Pirelli Foundation)

grooves, while other patterns featured diverse shapes.

Although documentation is quite sparse, it seems that at least initially the Superflex sigillo verde made of rayon (Raiflex) had a tread with grooves roughly in the shape of an 'H'.

In conjunction with the entry into force of the regulations concerning standardised military vehicles (issued in 1937, but implemented only over the next two to three years), a further variant of the sigillo verde appeared, called sigillo verde impero. In this tread the pattern had grooves in the shape of a lowercase 'f'. The sigillo verde impero pattern for Superflex and for Ultraflex tyres was highly employed in medium and heavy civilian trucks.

The other famous tread on tyres of Italian military vehicles, in particular tractors, armoured cars and vans, was the *artiglio* (claw), roughly comparable to the US NDT (non-directional tread) and the British 'cross-country' military tyres. From its launch in 1932, several variants were produced, outwardly similar to each other; there were, for example, semi-pneumatic and 'flex' artiglio tyres, with high and low inflating pressure. The artiglio tread was very sculpted and, according to the manufacturer, was suitable 'for heavy vehicles, intended for routes on sand, snow and marshy ground'; it was also designed to 'avoid the use of chains'. Actually, field experience showed that it tended to 'dig' into soft ground such as sand and mud until the vehicle bogged down; chains were still necessary in the snow.

For this reason, at the beginning of 1940 with the imminent opening of a war front in Africa, Pirelli put on the market a variant of the Superflex sigillo verde tread, for large-diameter wheels only (9.75 x 24 or 11.25 x 24 tyres – see below), such as to reduce the ground pressure, while decreasing the angle of incidence between the wheel and the sandy ground. The Superflex Libia, as it was called, had a great success even though production was never enough to equip all of the vehicles operating in North Africa.

The *stella bianca* (white star) tread for cars and light vehicles was the 'father' of the Sigillo Verde, having been conceived in 1927. The pattern of stella bianca for the Superflex family displayed characteristic 'Z' grooves, while the same for ultraflex was different and was also called *lusso* (luxury).

Particularly from 1942, due to the exhaustion of natural rubber stocks, tyres began to be manufactured in synthetic

The Superflex Sigillo Verde with the standard polygonal mesh pattern of tread.

The best known pattern for the Stella Bianca tread.

A Superflex Sigillo Verde Impero. Also in this tyre you can see 'Pirelli Superflex', the star, and '11.25 – 24'.

Two images of another Superflex Sigillo Verde, but with the tread pattern of the Raiflex variant. On the side you can read Pirelli Superflex and 11.25 – 24, with the star seal of Pirelli in the middle.

The Raiflex tread pattern for heavy vehicles featured 'H' shaped carvings, as in the case of the Cord 42 x 9 visible in section on the left. (Archivio Storico Fondazione Pirelli)

rubber, with a notable decline in quality and durability. However, after 8 September 1943, the Germans removed the machinery in use in Italy to continue production in Germany.

The two reference measures reported by the technical manuals are the nominal width of the tyre, and the diameter of the rim on which the tyre is mounted. However, the nomenclature was not unique.

In some cases, the two values were expressed in inches and separated by a dash, a slash or 'x' sign: 5.50 – 18 (but also 5.50 x 18) meant a 5½ inch width and a rim diameter of 18 inches.

In particular for the semi-pneumatic tyres, the two measures were instead expressed in millimetres, as for 175 x 720, that is a 175mm width and a rim diameter of 720mm.

There were, mainly in low and very low pressure tyres, mixed indications in millimetres and inches: 210 x 20 (but also 210 – 20) meant a 20-inch diameter rim and a 210mm width. Sometimes the two measures were written in reverse order.

Finally, it should be noted that usually the sidewall of the tyre had approximately the same size as the width.

Motorcycles had classic wire wheels. Special wire wheels were also fitted to some tractors. The wheels of the other motor vehicles were made of steel, essentially of two categories. The first category was made up of stamped discs, made of steel or other alloys, which could have circular or oval shaped lightening holes; often the front and rear wheels were interchangeable. Usually, the disc was bolted to the hub to facilitate disassembly.

The second category, which came into service in the second half of the 1930s and specifically for heavy vehicles, included sturdy wheels in cast steel characterised by six or eight large hollow spokes, forming a single body with the hub. The Gianetti company manufactured them in Italy having obtained the licence from the American Dayton Steel Foundry in 1932. In some texts this model of wheel is indicated with the denomination 'artillery

OVERVIEW AND EXPLANATORY NOTES 27

As you can read on its side, this tyre is a Pirelli Cord 32 x 6 artiglio early type.

A Superflex with artiglio tread. Note the different pattern compared with the 'Cord Artiglio'.

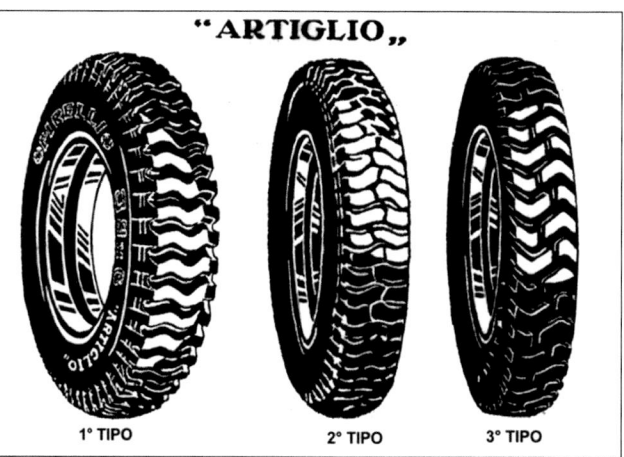

The artiglio tread pattern varied over time and depending on the tyre model. At least three types are recognised.

A Pirelli Superflex Libia 9.75 x 24, as can you read on its side.

type' inherited from the first wooden spoke wheels.

The rim, if not integral with the wheel disc, could be one-piece or divided in several parts. In particular, the Trilex model rim (born from a 1936 patent of the German company Georg Fischer) consisted of three segments fixed to the internal disc of the wheel with clamps; in twin wheels (double truck tyres) an intermediate external ring served as a spacer.

Semi-pneumatic tyres could be fitted to both categories of wheels but were preferably mounted on disc wheels. Cars and light trucks were equipped with disc wheels, with or without lightening holes. Trucks and tractors had their tyres usually mounted on spoked wheels.

The weakness of the tyres and the poor condition of the roads encountered in all operational theatres caused many punctures, for which there were not always enough spare tyres carried on the vehicles themselves, forcing repairs to be made continually, most of which did not last for long.

Electrical Systems

Even at the end of the thirties not all vehicles of the Italian armed forces were equipped with a battery, so the presence of a crank

A 20 inch spoked wheel Dayton-type manufactured by Gianetti. It consisted of a hexagonal-shaped cast steel sheet with six hollow spokes having holes at the ends for fixing to the rim; another six holes were present along the sides of the hexagonal disc (in this picture screws, bolts and other accessories are inserted). In the centre was the hub cap. (Museo delle Industrie e del Lavoro del Saronnese)

The effects of extended travel in the Libyan desert on the wheels of a Fiat 634. The outer tyre is a Pirelli Superflex with Sigillo Verde tread, the innermost with a much worn 'Sigillo Verde Impero' tread. (ACS)

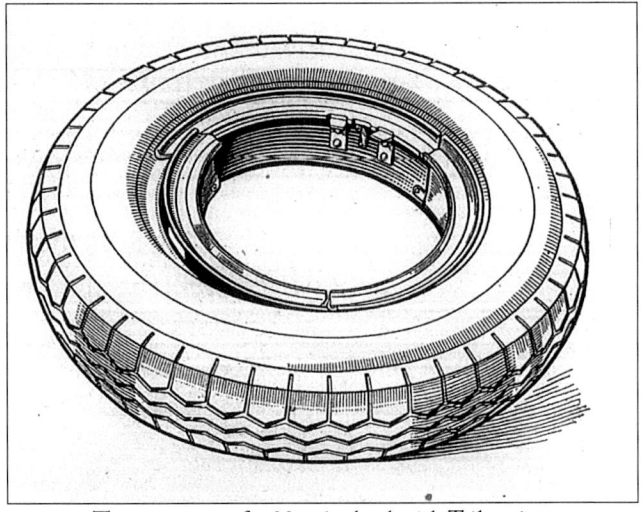

The structure of a 20 x 6 wheel with Trilex rims and tyres mounted on Fiat 626 trucks.

for starting the engine was an essential requirement. Since, as a result, the electric starter was also missing, an inertia starter could be present on heavy trucks and tractors; the case of the famous Lancia 3Ro is typical.

For the same reason, i.e. the lack of battery, other requirements for military trucks called for the presence of non-electric emergency headlights (petroleum, acetylene or carbide headlamps).

Right-Hand Drive

It might seem strange or puzzling to both British and American readers that Italian trucks were fitted with right-hand drive, in view of the fact that Italy practised the European custom of driving on the right-hand side of the road, although the legislation in this regard only entered force at the end of 1923 and became operational gradually.

The reason for this apparent anachronism lies in the fact that well into the 1920s and 1930s, outside of cities themselves, there were very few paved roads in Italy; this was also the reason that for many years the Italian Army requested that, in order to avoid punctures, trucks be fitted with semi-pneumatic rather than pneumatic tyres. Many roads were typically very narrow and shoulders were either nonexistent or were not very stable, neither were there any guard rails. The risk of ending up in a ditch or hitting a tree close to the side of the road was high; many trees close to the road had part of their trunks painted white precisely to act as a warning to drivers. Because of these reasons, having the steering wheel on the right-hand side enabled the driver to better judge how close the truck was to the edge of the road or to trees or other obstacles (large rocks, bridge parapets or, in some cases, even buildings) and thus avoid accidents. The steering wheel on the right-hand side of the vehicle made overtaking difficult, but heavy trucks were travelling at very low speeds and therefore overtaking other moving vehicles was infrequent. During the post-war years, Italian trucks transitioned to left-hand drive to conform to Continental driving practices.

Types of Bodywork

For the motor cars of the time, the two most common bodywork configurations, as for military purposes, were called in Italian 'torpedo' and 'berlina'.

The 'torpedo' body style, a tourer in England, was characterised by a completely open passenger compartment, with half-doors without windows, located immediately behind the engine compartment. The engine hood and the sides of the passenger compartment were equalised so as to present an almost continuous surface, we could say 'aerodynamic'. At the rear, the bodywork

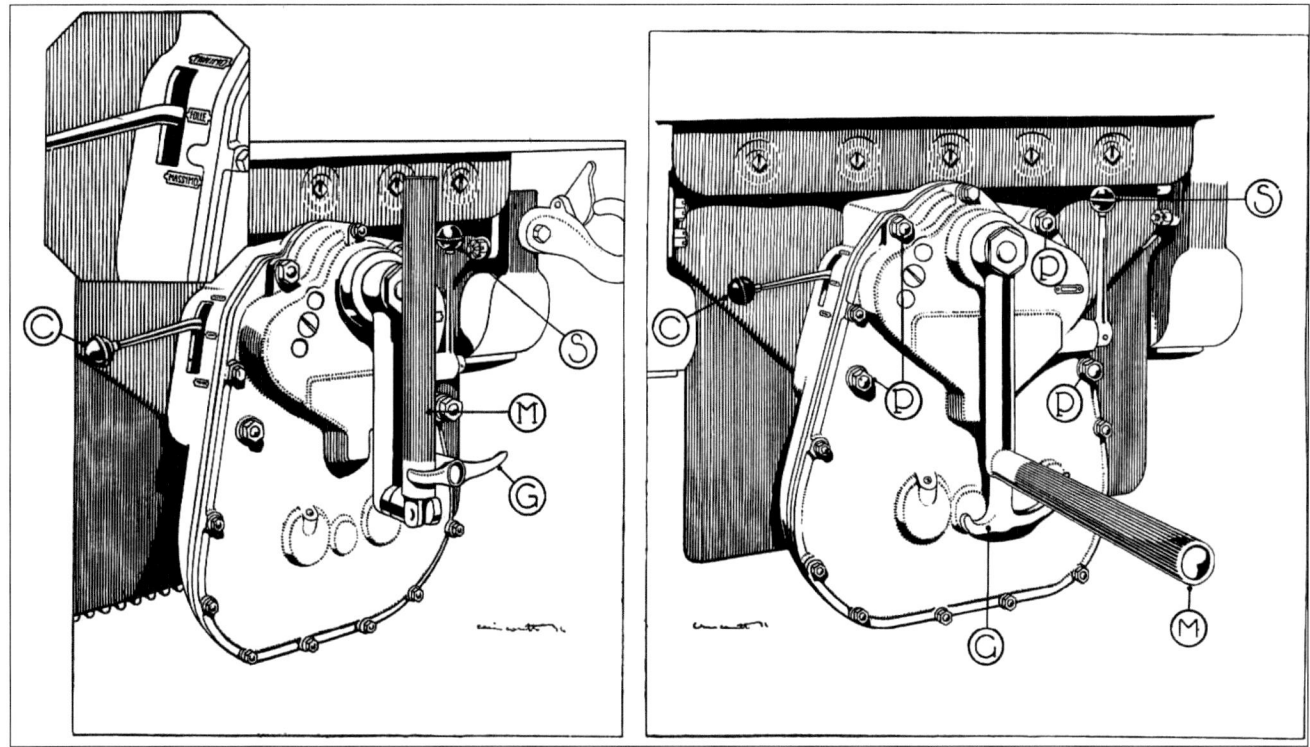

The inertia starter of the Lancia 3Ro truck.

ended with the passenger compartment itself, the tail being absent or at most limited to a small luggage compartment. There was usually a soft top, customarily in canvas, rigid or folding.

The 'berlina' body style, sedan or saloon in English, was and still is the classic closed body, with a fixed rigid roof, two-box or three-box design and, usually, with four or more doors.

To these two must be added the 'spider', the roadster in English, a sports body style for a two-seater, convertible car.

As far as heavy vehicles are concerned, the truck (otherwise called a lorry) was equipped with a rear bed, while the van has a completely closed body. It is appropriate to mention the types of cab. In the *cabina avanzata*, cab over engine (COE) in English, cab and engine are located above the front axle, therefore the front of the vehicle and the windshield are aligned, and the vehicle is 'flat nosed'. In the *cabina arretrata*, in the American cab – or conventional cab in England – the cab is located behind the engine and the nose of the vehicle is prominent. Finally, the *cabina semi-avanzata* is a compromise between the previous ones, in which the windshield is aligned with the front axle and the nose is small.

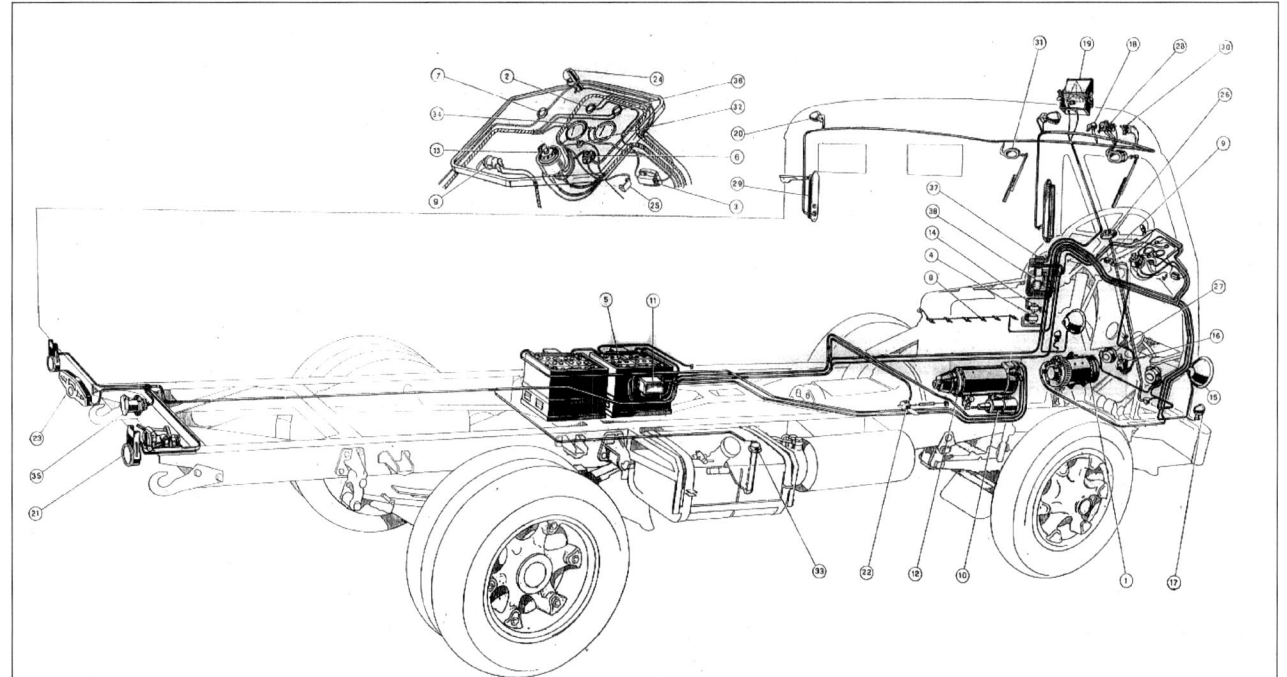

Electrical system of the Fiat 626 NLM truck.

2
Motorcycles

Motocicletta Volugrafo Aermoto

Despite the worn negative, the photo shows the overall structure of the small Aermoto.

German soldiers driving the Aermoto in Italy.

An example examined by British soldiers in Rome, in via Casilina, in June 1944.

Developmental and Service History

The firm Officine Meccaniche Volugrafo, based in Turin, produced fuel gauges for airplanes in the 1930s, which were also sold to the Luftwaffe. Volugrafo can in fact be translated as a volume indicator. It had already submitted some experimental models of vehicles to the Italian Army, without success.

The Volugrafo Aermoto Aviolanciabile was a parachutable light motorcycle based on a 120cc engine and designed in 1936 by engineer and company owner Claudio Belmondo. It was the forerunner of the post-war Vespa and Lambretta motor scooters, however, it was not itself successful following the war. Starting in 1938, the Italian armed forces were setting up the first airborne units and studying how to make paratroopers mobile once they were on the ground.

The Volugrafo Aermoto was accepted as a small parachutable motorcycle in conjunction with the planning for Operation C3, the invasion of Malta, which did not in fact occur. The final model was ready in 1942 and a first series of machines were produced in 1943. An initial issue of 2,000 examples was scheduled for the 183rd Paratroopers Division 'Ciclone', with Aermotos also being issued to the parachute school at Tarquinia and the 'San Marco' Regiment of the Italian Navy. However, the vehicle was never used en masse and many examples remained in the army's warehouses. Following the September 1943 Armistice, production of the Volugrafo Aermoto continued for a few months under German supervision and some examples were assigned to German paratroop units along the Adriatic coast and in the Rome area.

Technical Description

The Volugrafo Aermoto had many innovative features, especially for such a small machine. The frame was a rigid double cradle made of tubular steel, and exhaust gases were vented through the tubular frame itself, causing a lot of smoke and noise. The handlebars could be folded so that the machine could fit into a container for dropping by air; the specially designed container measured 1,300mm (51 inches) long by 800mm (31 inches) diameter and was cushioned internally with duck fabric. Once the container had landed, the Volugrafo Aermoto could be removed from the container, have its handlebars repositioned and locked, and be ready for action in two minutes. The gas tank was under the seat. The two-stroke engine was derived from the Sertum Batua 120 motorcycle running on a mix of petrol and oil. During the design phase, it was realised that the vehicle had to have wheels of a very small diameter but be capable of withstanding the weight of a man and his equipment; since suitable wheels and tyres did not exist, the choice fell on trolley wheels which were placed double on each axle; however, the motorcycle always tended to travel on the right or left wheels. After the war the Volugrafo Aermoto was modified and equipped with a single wheel. Brakes were drum brakes. The machine could tow a small trailer that had an 80kg (176lbs) carrying capacity.

Variants

A three-wheeled experimental variant, designated as a *motocarrello per aviotruppe* (motorised cart for airborne troops) was made but never put into full-scale production. It had a small cargo platform over the front wheel, while the operator sat behind it.

Specifications

- Designation: Motocicletta Volugrafo Aermoto
- Producer: Officine Meccaniche Volugrafo, Turin
- Years produced: 1942–1944
- Number produced: 600 (estimated)
- Length: 1,050mm (41.5 inches)
- Width: 310mm (12.2 inches) with handlebars folded
- Height: 530mm (20.87 inches)
- Unladen weight: empty 51.5kg (113.5lbs); equipped 59kg (130lbs)
- Tyres: 2.50 x 8
- Engine: Batua single cylinder, two-stroke, 120cc, 2 HP @3,600 rpm
- Transmission: two speeds, chain-driven
- Fuel capacity: 9.5 litres (2.5 US gallons, 2 Imperial gallons)
- Fuel: gasoline/oil mixture
- Speed: 50km/h (31 mph)
- Range (on road): 300km (186 miles)

Motociclo Benelli 250 M37

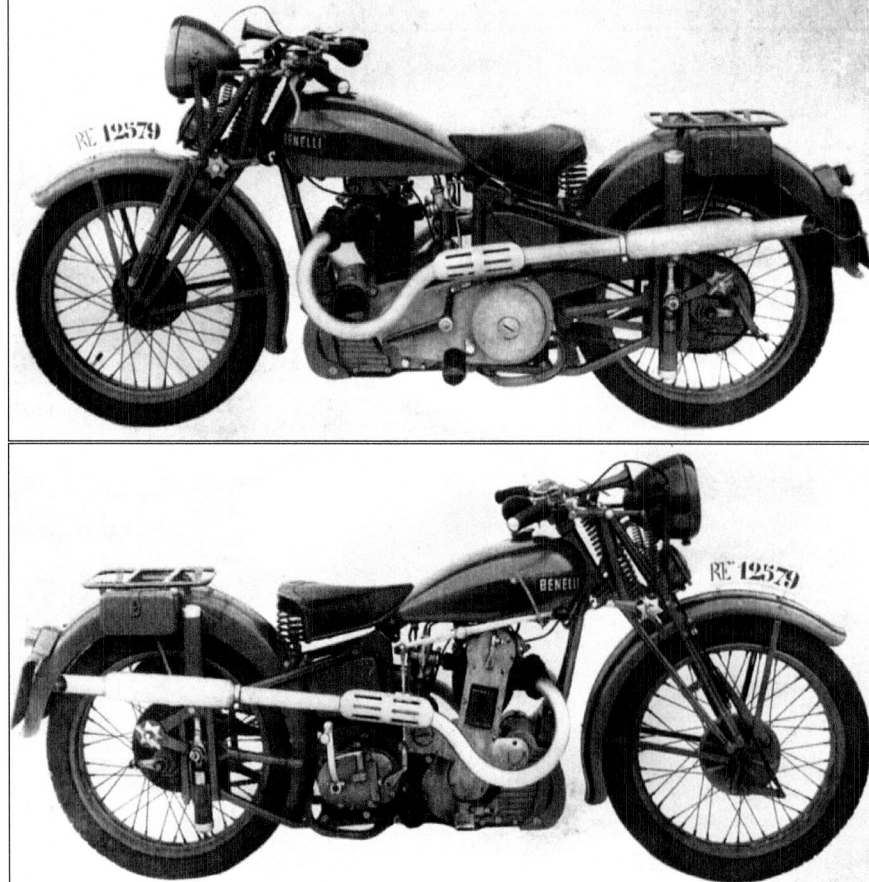

The Benelli 250 M37 was a military model derived from the civilian Benelli 250 TN. Of note are the two silencers, one on each side.

Developmental and Service History

The Benelli 250 M37 TE was a single-seat motorcycle whose production began in 1937; it was one of three types of Benelli motorcycle adopted by the *Regio Esercito* prior to the Second World War with which it shared many components, including engine, transmission, brakes and wheels. Motorcycles with 250cc engines like this were used in the army mainly for messenger and escort duties, leaving vehicles with 500cc engines (see for example the Benelli VL 500) at the disposal of the fighting troops. Details concerning numbers built and specific service history are lacking.

It should be noted that the Benelli 250 and almost all the other motorcycles mentioned in this work were also used by the Royal Carabinieri.

Technical Description

The Benelli 250 M37 TE (TE meant *militare mod. 1937 Telaio Elastico*, or military model 1937 with elastic frame) derived from the civilian 250 TN; it was a conventionally configured motorcycle with a single cylinder four-stroke engine inclined 12 degrees forward, manual transmission, a front fork with compression springs and adjustable dampers and a patented rear swing arm with shock absorbers. The fuel tank was above the engine, and the single seat was directly behind the fuel tank. The transmission was activated by a pedal and consisted of a primary gear and secondary chain. The front brake was activated by hand, the rear brake by a pedal. The electrical system was based on a Marelli dynamo.

Specifications

- Designation: Motociclo Monoposto Benelli 250 M37 TE
- Producer: Fabbrica Motoveicoli Fratelli Benelli, Pesaro
- Years produced: 1937–?
- Number produced: NA
- Length: 2,160mm (85 inches)
- Width: 730mm (28.74 inches)
- Height: 1,000mm (39.37 inches)
- Unladen weight: 147kg (324lbs)
- Unladen weight (with fuel, oil and tools): 161kg (355lbs)
- Wheelbase: 1397mm (55 inches)
- Tyres: Superflex 3.00 x 19
- Minimum turning radius: inclined vehicle 2,000mm (78.74 inches); upright vehicle 2,200mm (86.61 inches)
- Minimum clearance: 150mm (5.9 inches)
- Fording depth: 350mm (13.78 inches)
- Engine: Benelli 4TM single cylinder, four-stroke, air-cooled, 246.79cc, 9 HP @4,750 rpm
- Transmission: four speeds
- Fuel capacity: 12 litres (3.17 US gallons, 2.64 Imperial gallons)
- Speed: 66km/h (41 mph)
- Range (on road): 200km (125 miles)
- Range (cross-country): 5 hours

MOTOCICLO BENELLI 500 VL MILITARE

The Benelli 500 VL Militare, derived from the civilian model 500 VLC.

The Benelli 500 VL Militare, derived from the civilian model 500 VLC.

Developmental and Service history

The Benelli 500 VLM (*valvole laterali militare*, that is side valves, military) aka 500 VL TE M40 (*telaio elastico, militare modello 1940*; elastic frame model military 1940) derived from the civilian model 500 VLC. This motorcycle could be configured with either one or two seats. It shared many mechanical components with the smaller Benelli 250 M37.

Similarly to the Guzzi Alce 500, in the army the 500 VLM was employed for reconnaissance and liaison duties, for example equipping entire Bersaglieri (mobile light infantry) motorcycle battalions of the armoured and motorised divisions.

Details concerning numbers built and specific service history are lacking.

Technical Description

The Benelli 500 VLM was a conventionally configured motorcycle with a single cylinder four-stroke engine inclined 3 degrees forward, manual transmission, a front fork with compression springs and adjustable dampers and a patented rear swing arm with shock absorbers. The fuel tank was above the engine, and the seat of the single-seat version (*monoposto*) was directly behind the fuel tank, while that of the two-seat version (*biposto*) was over the rear fender. The transmission was activated by a pedal and consisted of a primary gear and secondary chain. The front brake was activated by hand, the rear brake by a pedal. The electrical system was based on a Marelli D30 dynamo, without battery.

Specifications

- Designation: Motociclo Militare Benelli 500 VL TE M40
- Producer: Fabbrica Motoveicoli Fratelli Benelli, Pesaro
- Years produced: 1940–?
- Number produced: Not known
- Length: 2,130mm (83.8 inches)
- Width: 800mm (31.5 inches)
- Height: 1,020mm (40.16 inches)
- Unladen weight: 178kg (392lbs) *monoposto*; 187kg (412lbs) *biposto*
- Wheelbase: 1,400mm (55.1 inches)
- Tyres: Superflex 3.50 x 19
- Minimum turning radius: 2,300 – 2,500mm (90.5 – 98.5 inches)
- Minimum clearance: 200mm (7.87 inches)
- Engine: Benelli single cylinder, four-stroke, air-cooled, 493.6cc, 11 HP @4,200 rpm
- Transmission: four speeds
- Fuel capacity: 12.5 litres (3.30 US gallons, 2.75 Imperial gallons)
- Speed: 86.5km/h (53.7 mph)
- Range (on road): 275km (171 miles) monoposto; 250km (155 miles) biposto

Mototriciclo Benelli 500 M36

The Benelli 500 M36 motor tricycle was the three-wheeled adaptation of the Benelli 500 VL motorcycle.

In view of the attack on France in 1940, a formation of Benelli 500 M36 equipped with a radio set R2 mod. 1935 (a radiotelegraphic station for infantry troops with an range of about 10km) was organised. (ACS)

A column of Benelli 500 M36 carrying cargo as well as personnel (Bersaglieri). Flanking the head of the column of trucks is a Fiat 1100 Camioncino.
(V. de Gaetano)

Developmental and Service History

The mototriciclo (motor tricycle) Benelli 500 M36 was the three-wheeled adaptation of the Benelli 500 VL motorcycle. Details concerning numbers built are missing.

The mototriciclo Benelli was requested in the mid-thirties by the *Regio Esercito* to motorise some infantry units, in particular the Bersaglieri who were considered 'rapid troops'. It began to be used in the Spanish Civil War, both for the transport of ammunition, light weapons and other materials, and the towing of 20mm Breda automatic cannons. Later, it was joined in all these roles by the Guzzi Trialce (see below).

Technical Description

The mototriciclo Benelli 500 was a three-wheeled motorcycle that shared many mechanical components of the Benelli 500 VL from which it derived. It had a single cylinder, four-stroke engine inclined 3 degrees forward, manual transmission and a front fork with compression springs and adjustable dampers; the rear swing arm and single wheel were replaced by a steel frame that held the two rear wheels with a suspended axle on leaf springs. The frame also held a wooden box body that had a 600kg carrying capacity. The fuel tank was above the engine, and the seat was directly behind the fuel tank. The transmission was activated by a pedal and consisted of a primary gear and secondary chain, acting on the rear differential gear. The front brake was activated by hand, the rear brakes by a pedal. The electrical system was based on a Marelli dynamo.

Specifications

- Designation: Mototriciclo Benelli 500 M36
- Producer: Fabbrica Motoveicoli Fratelli Benelli, Pesaro
- Years produced: 1936–?
- Number produced: NA
- Length: 3,020mm (118.9 inches)
- Width: 1,250mm (49.2 inches)
- Height: 1,020mm (40.16 inches)
- Unladen weight: 327kg without liquids (720.9lbs); 346kg (762.8lbs) in travelling order
- Wheelbase: 1,924mm (75.74 inches)
- Rear track: 900mm (35.43 inches)
- Tyres: Superflex 3.50 x 19 (old nomenclature 26 x 3.50)
- Minimum turning radius: 2,000 – 2,300mm (79 – 90.5 inches)
- Minimum ground clearance: 250mm (9.84 inches)
- Carrying capacity: 370kg (815lbs)
- Box body length: 1,240mm (48.82 inches)
- Box body width: 900mm (35.43 inches)
- Box body sides height: 320mm (12.6 inches)
- Engine: Benelli 4 TMN 500, single cylinder, four-stroke, air-cooled, 493.6cc, 12HP @4,500rpm
- Transmission: four speeds
- Fuel capacity: 13.5 litres (3.5 US gallons, 3 Imperial gallons)
- Speed: 72.6km/h (45 mph)
- Range (on road): 225 – 260km (140 – 162 miles)

Motociclo Bianchi 500 M

The Bianchi 500 M single-seat version (*monoposto*).

The Bianchi 500 M single-seat version (*monoposto*).

The Bianchi 500 M two-seat version (*biposto*).

Bersaglieri and a Bianchi 500 M with leg shields in North Africa in the spring of 1941. (BAMA)

A Bianchi 500 M single seat with standard Pirelli Superflex tyres. Note the *Regio Esercito* badge under the headlight.

The Bianchi motorcycle also operated on the Eastern front; shown here in 1942 with some Bianchi Miles trucks under repair. (L. Valente)

Developmental and Service History

In addition to supplying staff cars and trucks for Italian military forces during the Second World War, the Edoardo Bianchi company also supplied motorcycles that saw service in Italian East Africa and later in Spain. Beginning in 1936, Bianchi began to produce a new 500cc military motorcycle, the 500 M (*militare*, military) which operated throughout the war in Italy and in North Africa. The 500 M was produced in both a single-seat as well as a two-seat version. The Bianchi 500 M continued to be used by the Italian Army into the 1950s.

Technical Description

The Bianchi 500 M was a conventionally configured motorcycle with a single cylinder four-stroke engine mounted vertically, manual transmission, a front fork with compression springs and a rear swing arm. The fuel tank was above the engine, and the seat of the single seat version (*monoposto*) was directly behind the fuel tank, while that of the two-seat version (*biposto*) was over the rear fender. The transmission was activated by a pedal. The front brake was activated by hand, the rear brake by a pedal. The 6-volt electrical system used a Marelli D30 dynamo.

Specifications

- Designation: Motociclo Bianchi 500 M
- Producer: Edoardo Bianchi, Milan
- Years produced: 1936–1944
- Number produced: Not Known
- Length: 2,120mm (83.46 inches)
- Width: 750mm (29.5 inches)
- Height: 960mm (37.8 inches)
- Unladen weight: 170kg (375lbs) *monoposto*; 178kg (392lbs) *biposto*
- Wheelbase: 1,380mm (54.33 inches)
- Tyres: Superflex 3.50 x 19
- Minimum turning radius: 2,300mm (90.55 inches)
- Minimum clearance: 165mm (6.5 inches)
- Engine: Bianchi single cylinder, four-stroke, air-cooled, 498cc, 9 HP @3,200 rpm
- Transmission: three speeds
- Fuel capacity: 12 litres (3.2 US gallons, 2.6 Imperial gallons)
- Speed: 75km/h (47mph)
- Range (on road): 260km (162 miles) *monoposto*; 240km (149 miles) *biposto*
- Range (off-road): 200km (124 miles) *monoposto*

Motociclo Gilera 500 LTE

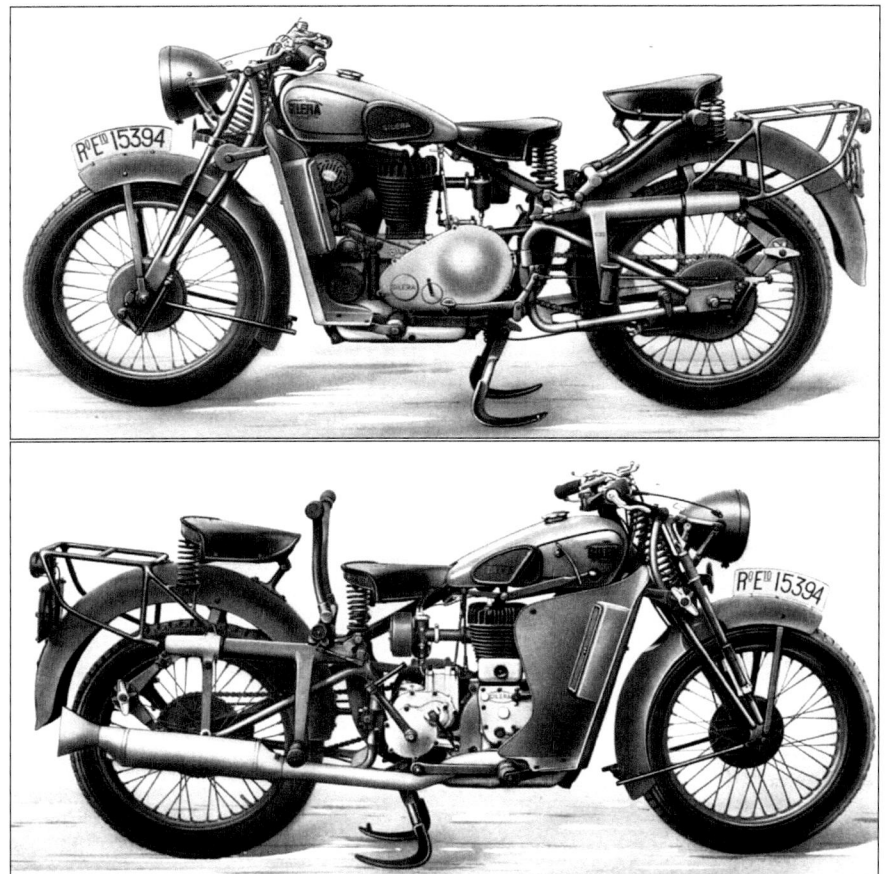

The Gilera 500 LTE.

Gilera 500 LTE with Breda 30 light machine gun.

An Italian column on the move in Egyptian territory. In the foreground, a Gilera 500 LTE in front of two Fiat 508 CMs. (ACS)

Spring 1941. Bersaglieri from the 7° *Reggimento* in the Tobruk area. (ACS)

Developmental and Service History

Gilera started its own production of motorcycles in 1909; its products had a reputation for having exceptional performance. In 1937 the company began manufacturing a 500cc motorcycle for the *Regio Esercito*. It was produced in single-seat and two-seat configurations as well as in a three-wheeled version. The 500 LTE (originally VLTE, *Valvole Laterali, Telaio Elastico*, that is side valves, Elastic Frame) served on all fronts during the Second World War and remained in production until 1944, as well as serving in the post-war Italian Army for a while. Production figures are unknown.

Technical Description

The Gilera 500 LTE was a motorcycle powered by a vertically mounted single cylinder four-stroke engine. The tubular steel frame consisted of a parallelogram front fork and a patented flexible rear arm. The four-speed transmission was operated by a lever to the right of the gas tank; the LTE had a chain drive and had drum brakes. The fuel tank was above the engine, and the seat of the single seat version (*monoposto*) was directly behind the fuel tank, while that of the two-seat version (*biposto*) was over the rear fender; there was a set of dummy handlebars between the front and rear seats for the passenger to hold on to. The 6-volt electrical system was based on a Marelli D30 dynamo.

Specifications

- Designation: Motociclo Gilera 500 LTE Militare
- Producer: Gilera, Arcore
- Years produced: 1937–1944
- Number produced: Not Known
- Length: 2,200mm (86.61 inches) *monoposto*; 2,260mm (88.98 inches) *biposto*
- Width: 800mm (31.5 inches)
- Height: 1,050mm (41.34 inches)
- Unladen weight: 190kg (419lbs) *monoposto*; 203kg (448lbs) *biposto*
- Wheelbase: 1,450mm (57 inches) *monoposto*; 1,500mm (59 inches) *biposto*
- Tyres: Superflex 3.50 x 19
- Minimum turning radius: inclined vehicle 1,400mm (55.12 inches); upright vehicle 1,600mm (63 inches)
- Minimum clearance: 140mm (5.51 inches)
- Engine: Gilera 500 L single cylinder, four-stroke, air-cooled, 498cc, 12HP @3,800 rpm
- Transmission: four speeds
- Fuel capacity: 12 litres (3.2 US gallons, 2.6 Imperial gallons)
- Speed: 76km/h (47mph), later 80km/h (49mph) *monoposto*; 80km/h (49mph) *biposto*
- Range (on road): 230km (143 miles) *monoposto*; 220km (137 miles) *biposto*

Motocarrozzetta Gilera Marte 500

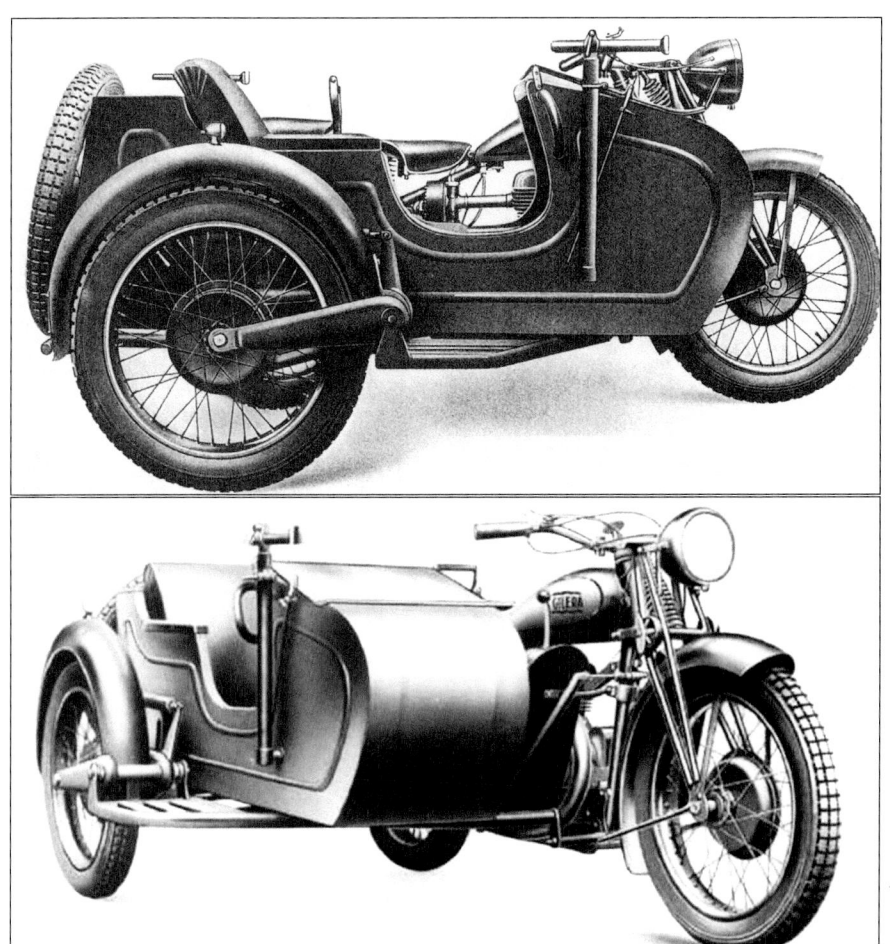

The Gilera Marte 500 sidecar.

Rear view of the Marte 500.

Developmental and Service History

At the end of 1940, Gilera designed the Marte 500, referred to as a *motocarrozzetta* (sidecar) based on the Gilera LTE 500 motorcycle (see above). The *Marte* (Mars) is said to have been inspired by the German sidecar configuration. It was developed to replace motorcycles equipped with light machine guns mounted on the handlebars in the motorcycle platoons and served throughout the war. After the Armistice it was tested by the *Oberkommando Wehrmacht*, which highlighted its defects, namely low power and high weight. At the end of the war, 158 that were completed and stored in a warehouse were converted to a civilian configuration. Total production figures are unavailable.

Technical Description

The Gilera Marte 500 was a motorcycle powered by a vertically mounted single cylinder four-stroke engine. A Cardan drive transmitted power to the rear wheel of the motorcycle, while a second shaft to the sidecar's wheel could be disengaged while the Marte was travelling on the road. The sidecar's wheel could oscillate freely on its own fork. The sidecar body had a mount on which a Breda model 30 light machine gun was fixed; a spare wheel with its tyre was mounted on the rear of the sidecar.

Specifications

- Designation: Motocarrozzetta Gilera Marte 500
- Producer: Gilera, Arcore
- Years produced: 1941–1945
- Number produced: Not Known
- Length: 2,300mm (90.55 inches)
- Width: 1,600mm (63 inches)
- Height: 1,020mm (40.15 inches)
- Unladen weight: 300kg (661lbs)
- Wheelbase: 1,400mm (55.11 inches)
- Rear track: 1,500mm (59 inches)
- Tyres: Superflex 3.50 x 19
- Minimum clearance: 180mm (7 inches)
- Minimum turning radius: 5,000mm (197 inches)
- Carrying capacity: 240kg (529lbs)
- Engine: Gilera 500 L single cylinder, four-stroke, air-cooled, 498cc, 14 HP @4,800 rpm
- Transmission: four speeds
- Fuel capacity: 16 litres (4.2 US gallons, 3.5 Imperial gallons)
- Speed: 78km/h (48 mph)
- Range (on road): 220km (137 miles)
- Range (cross-country): 180km (112 miles)

Motocarro Gilera Mercurio

The Gilera Mercurio motor tricycle with tarpaulin cover. Drawing from the user manual.

The Gilera Mercurio without tarpaulin.

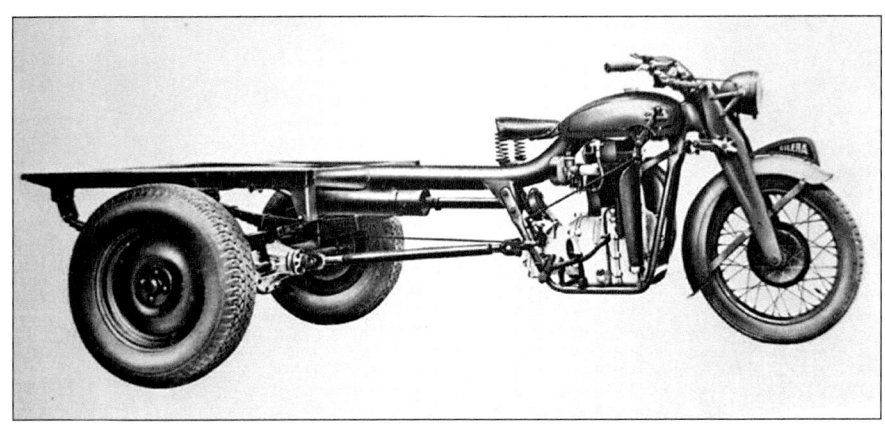

The frame of the Mercurio.

Developmental and Service History

The Gilera Mercurio was a three-wheeled motorcycle designed as *motocarro* (three-wheeler light truck) and based on the Gilera 500 LTE motorcycle. Gilera developed the *Mercurio* (Mercury) in 1938, but production was delayed until 1940. The Mercurio served on all fronts during the war. The September 1943 Armistice put an end to production; some or most of the Mercurios in the *Regio Esercito* inventory were then appropriated by Mussolini's *Esercito Nazionale Repubblicano* (National Republican Army) in northern Italy. After the war, production of a civilian version of the Mercurio continued until 1963.

Technical Description

The Gilera Mercurio shared the same mechanical components as the 500 LTE motorcycle from which it derived (see above). It had a single cylinder four-stroke engine; its transmission had four speeds forward as well as one reverse speed; power to the rear wheels was by a Cardan drive. The single rear wheel was replaced by a set of two wheels mounted on a frame behind the driver's seat; the frame held a small wooden cargo body with raised sides and tailgate which could be covered by a canvas cover supported by three bows. The normal motorcycle wire-spoke wheels were replaced by all-metal pressed steel wheels in the rear, while the wire-spoke wheel remained as the front wheel. The Mercurio had drum brakes; the rear brakes could be locked to stop the vehicle on sloping roads. The 6 V electrical system used a Marelli D30 dynamo.

Specifications

- Designation: Motocarro Gilera Mercurio
- Producer: Gilera, Arcore
- Years produced: 1940–1963
- Number produced: approximately 1,000 produced during the Second World War; production of a civilian version continued until 1963
- Length: 3,560mm (140.15 inches) standard type (short); 1,600mm (148.03 inches) lengthened type
- Width: 1,500mm (59 inches) standard type (short); 1,600mm (62.99 inches) lengthened type
- Height: 1,400mm (55.11 inches)
- Unladen weight: 580kg (1,278lbs) standard type (short); 610kg (1,385lbs) lengthened type
- Wheelbase: 2,230mm (87.8 inches) standard type (short); 2,430mm (95.67 inches) lengthened type
- Rear track: 1,260mm (49.6 inches)
- Tyres: Superflex 3.50 x 19 front; 6.00 x 16 rear
- Minimum clearance: 195mm (7.68 inches)

- Minimum turning radius: 3,900mm (154 inches)
- Carrying capacity: 1,500kg (3,307lbs); 1,000kg (2,204lbs) standard type.
- Box body length: 2,020mm (79.52 inches) standard type (short); 2,430mm (95.67 inches) lengthened type
- Box body width: 1,420mm (55.9 inches) standard type (short); 1,550mm (61.02 inches) lengthened type
- Box body sides height: 390mm (15.35 inches) standard type; 405mm (15.95 inches) lengthened type
- Engine: Gilera 500 L single cylinder, four-stroke, air-cooled, 498cc, 18 HP @4,100 rpm
- Transmission: four speeds forward, one reverse
- Fuel capacity: 15 litres (4 US gallons, 3.3 Imperial gallons)
- Speed: 70km/h (43mph)
- Range (on road): 200km (124 miles)

Motociclo Sertum 500 MCM

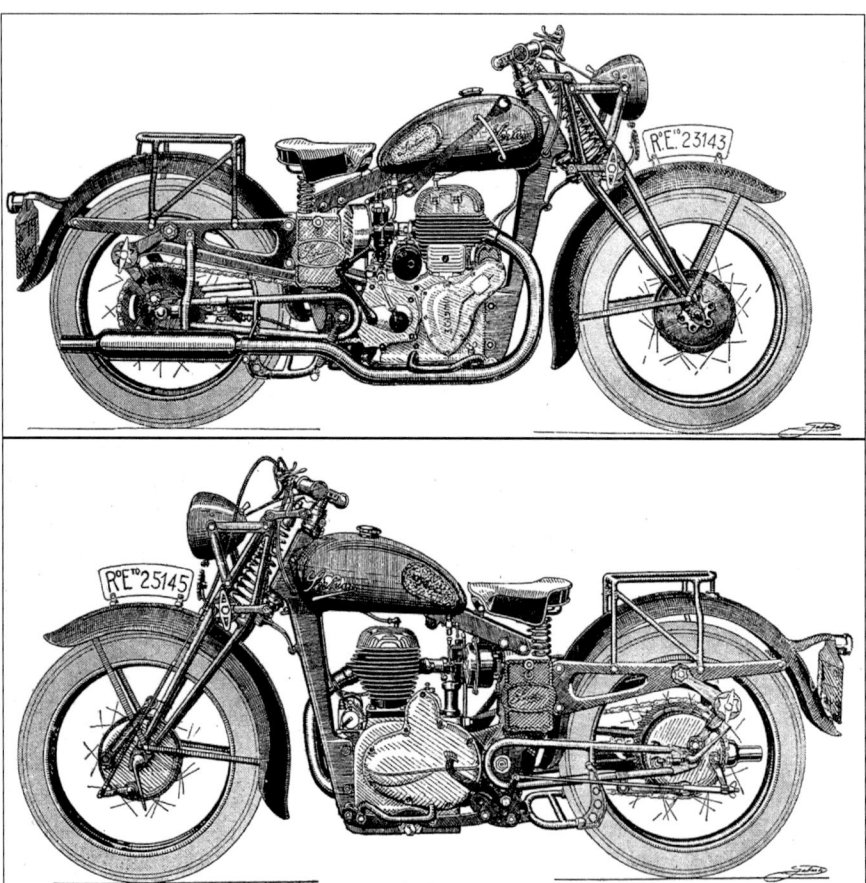

The Sertum 500 MCM single-seat.

The Sertum 500 MCM two-seat.

The Sertum 500 MCM two-seat.

The three-wheeled Motocarro 500 MCM.

A Sertum 500 MCM single-seat with *Regio Esercito* number plate. (ACS)

Developmental and Service History

The firm of Fausto Alberti of Milan, specialised in industrial engines and mechanical parts, was a relative newcomer to the motorcycle field, not having been established until 1932, marketing its machines under the brand name of Sertum (the Latin word for wreath). By 1937 Sertum had achieved a status equal to that of the more established names of Benelli, Bianchi, Gilera and Moto Guzzi. With the advent of the Second World War, Sertum developed a 500cc military machine in both single-seat (*monoposto*) and two-seat (*biposto*) versions, unveiling its 500 MCM in 1941. The vehicle derived from the civilian version 500

VL (*Valvole Laterali*). Subjected to rigorous testing, it was received enthusiastically by the military authorities. It was considered to be the most modern motorcycle assigned to motorcycle units in the Greek and North African campaigns. It remained in service with the Italian Army until the mid–1950s.

Technical Description

The Sertum 500 MCM was a conventionally configured motorcycle with a single cylinder four-stroke engine with side valves mounted vertically. The suspension's front fork used central compression coil springs; the rear suspension consisted of a swing arm with a leaf spring located under the seat. The fuel tank was above the engine, and the seat of the single seat version was directly behind the fuel tank, while that of the two-seat version was over the rear fender; there was a set of foldable handlebars between the front and rear seats for the passenger to hold on to. The transmission was activated by a pedal. The front drum brake was activated by hand, the rear drum brake by a pedal. The 6 V electrical system used a Marelli D30 dynamo.

Variants

A three-wheeled version, the Motocarro 500 MCM, was produced. It was similar to the Mototriciclo Benelli 500 M36 and the Guzzi Trialce (see separate entries). Its wooden box body measured externally 2,020mm (6ft 7in) x 1,420mm (4ft 8in) x 350mm (1ft 2in); it had a 500kg (1,102lbs) maximum carrying capacity.

Specifications

- Designation: Motociclo Sertum 500 MCM
- Producer: Officine Meccaniche Fausto Alberti, Milan
- Years produced: 1941–1946
- Number produced: Not Known
- Length: 2,190mm (86.22 inches)
- Width: 850mm (33.46 inches)
- Height: 1,070mm (42.12 inches)
- Unladen weight: 104kg (229lbs) *monoposto*; 113kg (249lbs) *biposto*
- Unladen weight (with fuel, oil and tools): 177kg (390lbs) *monoposto*; 186kg (410lbs) *biposto*
- Wheelbase: 1,450mm (57 inches)
- Tyres: Superflex 3.50 x 19
- Minimum clearance: 170mm (6.69 inches)
- Engine: Alberti Tipo 500 MCM single cylinder, four-stroke, air-cooled, 498cc, 12 HP @4,100 rpm
- Transmission: four speeds
- Fuel capacity: 13 litres (3.43 US gallons, 2.85 Imperial gallons)
- Speed: 77km/h (48 mph) *monoposto*; 73km/h (45 mph) *biposto*
- Range (on road): 200km (125 miles)

Motociclo Guzzi 500 Alce

The Guzzi 500 Alce single-seat. Note the characteristic double silencer on the left side. (Moto Guzzi)

The Guzzi 500 Alce two-seat. (Moto Guzzi)

A *biposto* (two-seat) seen from the right side. (Moto Guzzi)

The sidecar version. (Moto Guzzi)

Developmental and Service History

The Guzzi 500 *Alce* (elk, moose) stemmed from a number of earlier civilian Guzzi motorcycles since 1928, beginning with the GT, then followed by the Sport 14 in 1930, then by the GT 17 in 1932, GTV in 1934, S in 1937, and in 1939 by the transitional GT 20. The Guzzi 500 Alce – as well as the previous models – was adopted by the *Regio Esercito* and, starting from 1940, was widely issued to the three Italian armed forces and to the *Polizia dell'Africa Italiana*. In the army it was assigned to each infantry regiment for reconnaissance and liaison duties, and in the armoured and motorised divisions in North Africa it equipped entire Bersaglieri motorcycle battalions.

The Alce was issued in three versions: a single-seat (*monoposto*), a two-seat (*biposto*), and a version with a removable sidecar (*motocarrozzetta*). Some Alces were fitted with a special support for a 6.5mm Breda model 30 light machine gun; the gun could

The heavy machine gun carrier version. (Moto Guzzi)

Bersaglieri, riding Guzzi motorcycles, about to cross a rough wooden bridge as Rumanian locals look on in the background. (AUSSME)

not be used while the machine was moving, but the support enabled it to be used while stationary. Another variant included the 8mm Breda model 37 machine gun, which was heavier than the model 30, so it could only be carried hanging on the left side, while the tripod was hinged on the right side. Production of the Alce continued throughout the war until 1945.

Technical Description

The Guzzi 500 Alce was a conventionally configured motorcycle which had a horizontal four-stroke engine, manual transmission, a parallelogram front fork with friction dampers and a rear fork likewise with friction dampers; dampers were made by means of discs of friction material (such as cork) tightened between metal discs clamped together. The fuel tank was above the engine, and the seat of the single-seat version was directly behind the fuel tank; the second seat of the two-seat version was slightly raised above the rear fender sand had a separate set of foldable handlebars for the passenger to hold onto. The transmission consisted of a primary gear and secondary chain. The front brake was activated by hand, the rear brake by a pedal. The rear wheel had a detachable hub, i.e. brake and crown were placed on the same side, so that the disassembly of the chain to remove the wheel was

The heavy machine gun carrier version. (Moto Guzzi)

Bersaglieri, riding Guzzi motorcycles, about to cross a rough wooden bridge as Rumanian locals look on in the background. (AUSSME)

not be used while the machine was moving, but the support enabled it to be used while stationary. Another variant included the 8mm Breda model 37 machine gun, which was heavier than the model 30, so it could only be carried hanging on the left side, while the tripod was hinged on the right side. Production of the Alce continued throughout the war until 1945.

Technical Description

The Guzzi 500 Alce was a conventionally configured motorcycle which had a horizontal four-stroke engine, manual transmission, a parallelogram front fork with friction dampers and a rear fork likewise with friction dampers; dampers were made by means of discs of friction material (such as cork) tightened between metal discs clamped together. The fuel tank was above the engine, and the seat of the single-seat version was directly behind the fuel tank; the second seat of the two-seat version was slightly raised above the rear fender and had a separate set of foldable handlebars for the passenger to hold onto. The transmission consisted of a primary gear and secondary chain. The front brake was activated by hand, the rear brake by a pedal. The rear wheel had a detachable hub, i.e. brake and crown were placed on the same side, so that the disassembly of the chain to remove the wheel was

The Guzzi 500 Alce two-seat. (Moto Guzzi)

A *biposto* (two-seat) seen from the right side. (Moto Guzzi)

The sidecar version. (Moto Guzzi)

Developmental and Service History

The Guzzi 500 *Alce* (elk, moose) stemmed from a number of earlier civilian Guzzi motorcycles since 1928, beginning with the GT, then followed by the Sport 14 in 1930, then by the GT 17 in 1932, GTV in 1934, S in 1937, and in 1939 by the transitional GT 20. The Guzzi 500 Alce – as well as the previous models – was adopted by the *Regio Esercito* and, starting from 1940, was widely issued to the three Italian armed forces and to the *Polizia dell'Africa Italiana*. In the army it was assigned to each infantry regiment for reconnaissance and liaison duties, and in the armoured and motorised divisions in North Africa it equipped entire Bersaglieri motorcycle battalions.

The Alce was issued in three versions: a single-seat (*monoposto*), a two-seat (*biposto*), and a version with a removable sidecar (*motocarrozzetta*). Some Alces were fitted with a special support for a 6.5mm Breda model 30 light machine gun; the gun could

Bersaglieri intent on overcoming a muddy stretch with an Alce. (ACS)

A column of Guzzi GT 17 two-seat, the 'grandmother' of the Guzzi Alce. (AUSSME)

A column of Guzzi motorcycles carrying Bersaglieri in a Ukrainian village; the lead is a Guzzi GT 17 *monoposto* and is fitted with a Breda Model 1930 light machine gun, while most of the other Guzzis in the group are the GT 17 biposto version. (ACS)

not necessary; the operation could be carried out with the motorcycle hoisted on the central stand. Like almost all motorcycles of the time, it was equipped with a 6 V electrical system with Marelli D30 dynamo. It could be equipped with leg guards.

Variants
1,741 examples of a three-wheeled version, the Guzzi Trialce, were also produced (see below).

Specifications
- Designation: Motociclo Guzzi Alce
- Producer: Moto Guzzi, Mandello del Lario
- Years produced: 1940–1945
- Number produced: 6,390 plus 669 with sidecar
- Length: 2,220mm (87.4 inches)
- Width: 790mm (31.1 inches) *monoposto* and *biposto*; 1,575mm (62 inches) sidecar
- Height: 1,065mm (41.93 inches)
- Unladen weight (with fuel, oil and tools): 179.5kg (398lbs) *monoposto*, 187kg (412lbs) *biposto*; 260kg (573lbs) sidecar
- Wheelbase: 1,455mm (57.28 inches)
- Tyres: Superflex 3.50 x 19

Moto Guzzi GT 17s with Breda 30 machine gun.

Another Moto Guzzi GT 17 with Breda 30 machine gun.

- Minimum clearance: 210mm (8.26 inches)
- Engine: Guzzi single cylinder air-cooled, 498.4cc, 13.2 HP @4,000 rpm
- Transmission: four speeds
- Fuel capacity: 13.5 litres (3.5 US gallons, 3 Imperial gallons)
- Speed: 90km/h (56mph) *monoposto*; 85km/h (52mph) *biposto*
- Range (on road): 300km (186 miles)

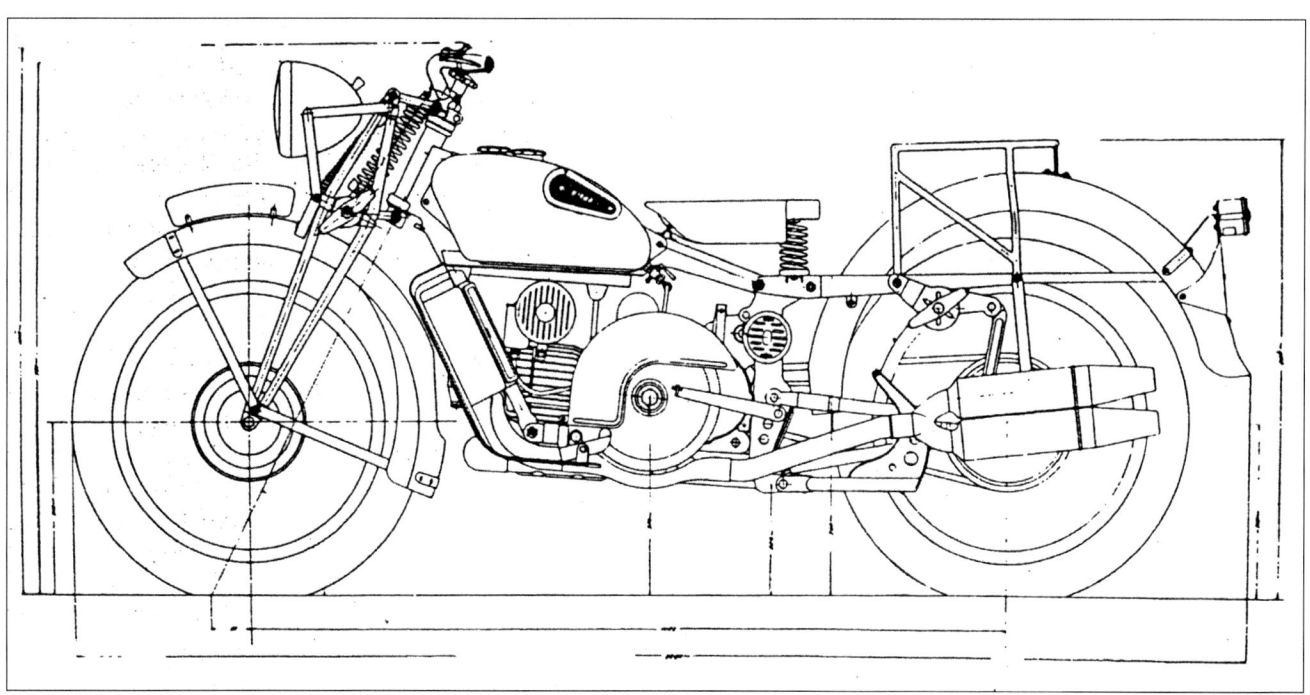

Line drawing of the Alce *monoposto* (left side view). (Guzzi)

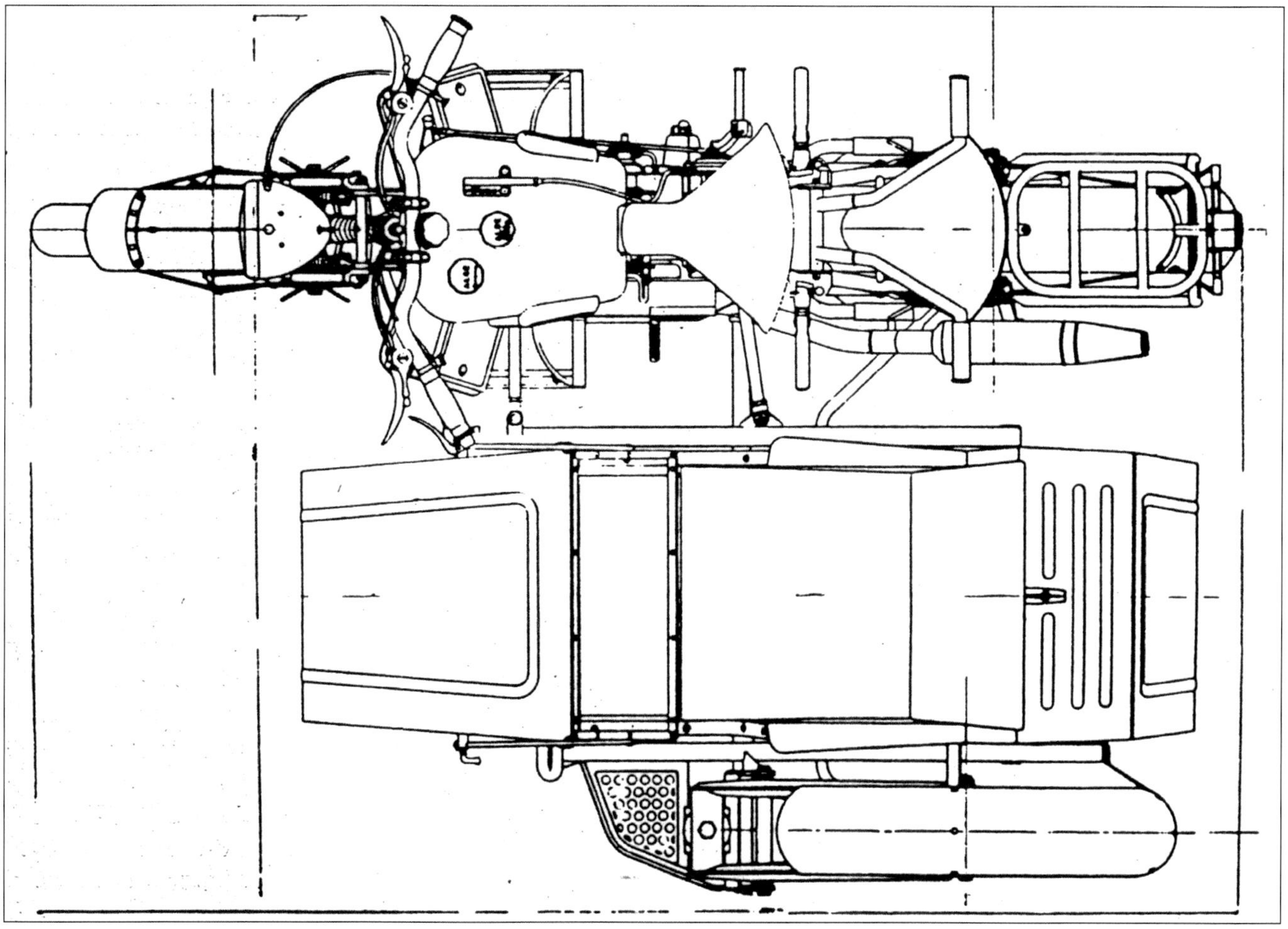

Line drawing (top view) of the Alce Motocarrozzetta. (Guzzi)

Mototriciclo Guzzi Trialce

The Guzzi Trialce motor tricycle. (Moto Guzzi)

Front and rear view of the Trialce. (Moto Guzzi)

Developmental and Service History

During the thirties, the Guzzi firm had entered the market of special three-wheeled motorcycles for the transport of materials, called *mototricicli* (motor tricycles). In 1940 the most modern model was launched, called Trialce and based on the proven chassis of the 500 Alce. 1,741 examples were produced until production ceased in 1943.

Beyond general uses, from the end of 1941 the Guzzi Trialce equipped the so-called 'motorcycle-machine-gun companies', who were also equipped with single-seat and two-seat motorbikes as well as trucks. Each Trialce carried, in addition to the machine gun, the gunner, the assistant and the ammunition boxes.

A Bersagliere riding a Guzzi Trialce past the Arco dei Fileni (Arch of the Philaeni), which marked the boundary between the regions of Tripolitania and Cyrenaica in Libya.

Guzzi motor tricycles being used to tow 65/17 guns of the 'La Spezia' Division's *80° Reggimento Artiglieria* in Tunisia, early 1943. (E. Finazzer)

A Guzzi Trialce in the Ukraine being used to transport Russian prisoners. (AUSSME)

The Mototriciclo Guzzi modello 32, manufactured between 1932 and 1939, was the predecessor of the Trialce. Photographed here in use in North Africa.

Technical Description

The Trialce used the front chassis, the fork and the engine of the Alce, a single cylinder four-stroke. The primary transmission was made of helical gears, the secondary used a roller chain, with a four-speed gearbox. The rear of the structure was modified with the installation of a suspension axle frame on helical springs and chain drive on the central differential. The frame supported a wooden bed. The wheels were interchangeable. The brakes remained drum brakes, the front ones operated by hand and the rear ones by pedal.

Another ancestor of the Trialce was the Guzzi ER manufactured from 1938. Here an example belonging to the *Unione Nazionale Protezione Antiaerea* (National Organisation for Anti-aircraft Protection); UNPA employed civilian personnel, who were mostly female.

Nice close-up of a captured Trialce. The writing relating to tyre inflation pressure stands out against the grey-green colour.

Variants

A variant of the Trialce could carry, in addition to the personnel, a Breda 37 machine gun on an anti-aircraft carriage, plus a tripod for ground fire.

The most original version was undoubtedly the air-transportable version assigned to the paratroops that could be disassembled.

Specifications

- Designation: Mototriciclo Guzzi Alce
- Producer: Moto Guzzi, Mandello del Lario
- Years produced: 1940–1943
- Number produced: 1,741
- Length: 2,825mm (9ft 3in)
- Width: 1,240mm (4ft 1in)
- Height: 1,050mm (3ft 5in)
- Unladen weight (with fuel, oil and tools): 354kg (780lbs)
- Wheelbase: 1,880mm (6ft 2in)
- Tyres: Superflex 3.50 x 19
- Minimum clearance: 210mm (8.26 inches)
- Carrying capacity: 400kg (881.85lbs) normal; 500kg (1,102lbs) maximum
- Box body length: 1,300mm (4ft 3in) external; 1,250mm (4ft 2in) internal
- Box body width: 950mm (3ft 1in) external; 900mm (2ft 11in) internal
- Box body sides height: 350mm (1ft 1in) external; 315mm (1ft) internal
- Engine: Guzzi single cylinder air-cooled, 498.4cc, 13.2 HP @4,000 rpm
- Transmission: four speeds
- Fuel capacity: 16 litres (4.2 US gallons, 3.5 Imperial gallons)
- Speed: 73km/h (45mph)
- Range (on road): 260km (161 miles)

3
Motor Cars

Autovettura Alfa Romeo 2500 C

The civilian Alfa Romeo 2500 6C modified for testing in East Africa. The tools behind the trunk and the additional fuel tanks located at the base of the rear mudguards and next to the spare wheels can all be clearly seen.

A 2500 in Eritrea in 1939, more precisely in Asmara and the Cheren (Keren) area. It bears the regulation 'Prova' number plate for vehicles not yet registered to be subjected to tests and trials. Tyres are Aerflex Stella Bianca. (G.B. Guidotti)

Developmental and Service History
In 1938 the Minister of War set forth a request for a colonial version of the Alfa Romeo 2500 6C (Turismo and Sport versions), to be used as staff car for higher-ranking officers and national authorities. Two examples were ready by 1939, and in October of that year began field trials in Eritrea and Ethiopia. The cars were subjected to a great deal of cross-country travel, as well as periods of travel on improved roads. In particular, the engineers wanted to find a correct setting of the carburettor at high altitudes and to check the oil cooling system. The first ever use of the cars took place in the Addis Ababa area (elevation over 2,400 m).

A 2500 in Eritrea in 1939, more precisely in Asmara and the Cheren (Keren) area. It bears the regulation 'Prova' number plate for vehicles not yet registered to be subjected to tests and trials. Tyres are Aerflex Stella Bianca. (G.B. Guidotti)

The pre-series 2500 C (Coloniale) with camouflage matt paint.

The 2500 C series version. The additional rear tanks are of the type, incorporated into the bodywork so as not to break the profile.

General Gastone Gambara, Chief of Staff of the armed forces Higher Command in North Africa, here in the region of Cyrenaica in 1941.

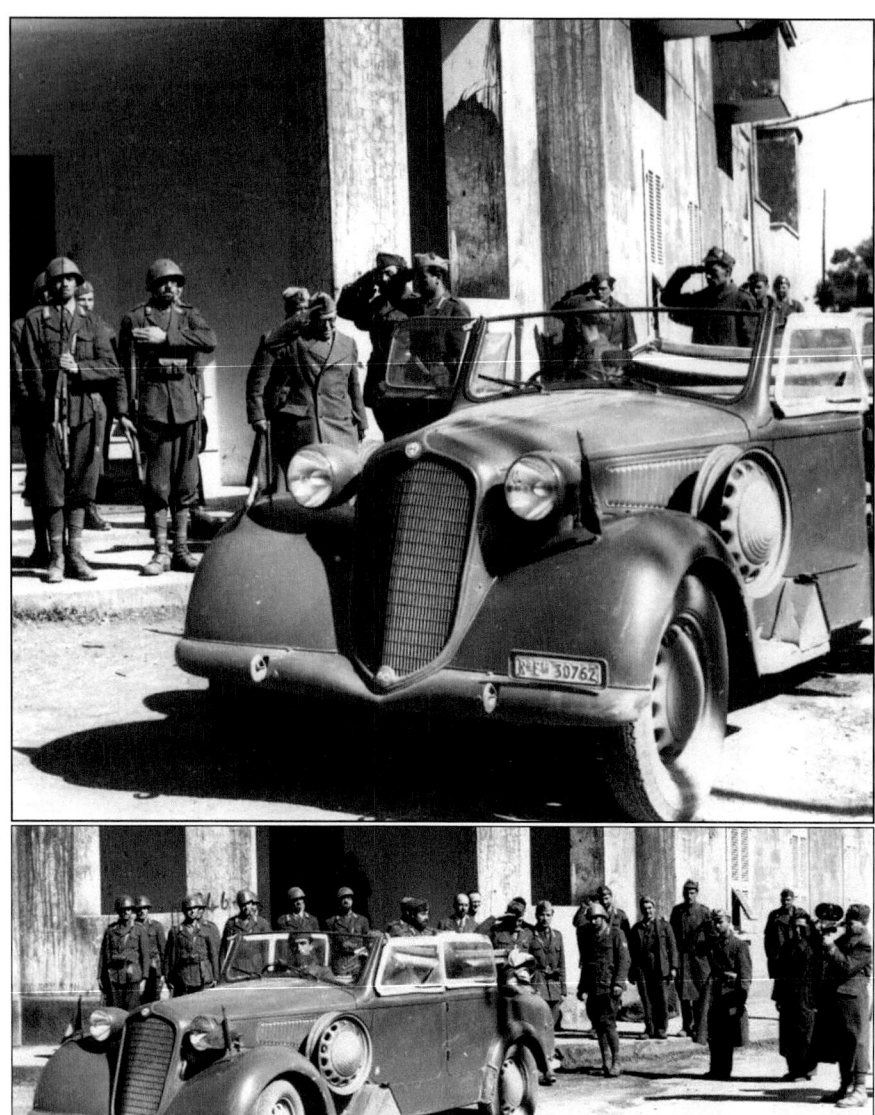

General Francesco Zingales, commander of XX Corpo d'Armata (20th Army Corps), visits the command of the 6th Grenadier Battalion in Libya in the winter 1941–1942. (ACS)

The trials led to some modifications and more tests, and the definitive 2500 C (Coloniale) entered service in 1941. In North Africa the car was less than completely satisfactory, receiving complaints regarding its suspension and tyres, unlike service in Russia (closed cab variant) where it was judged to be suitable for long trips over unimproved roads, although there was some dissatisfaction with its suspension in Russia as well. Among others, the 2500 C was used by Generals Gastone Gambara and Erwin Rommel in North Africa, by Prince Umberto of Savoy and by Mussolini himself. Production continued until the end of war resulting in a total of 194 machines (150 for the *Regio Esercito* in 1941 and 1942, plus 44 for the occupying German forces in 1944 and 1945), plus the two prototypes.

Technical Description
The civilian Alfa Romeo 2500 6C (six cylinders) was a luxury car by the standards of the day, and the derivative 'colonial' version 2500 C was an equally impressive car.

The Alfa 2500 C was a four-door open touring car (torpedo, according to the period nomenclature) with a retractable canvas roof and had a dividing partition between the driver's compartment and the passenger compartment; the bodywork was made by the coachbuilder Castagna. It was a 4x2 car with rear wheel drive and right-hand driven. Pneumatic tyres were mounted on pressed steel wheels with 14 holes around the perimeter of the rim; two spare wheels and tyres were mounted between the front doors and the front fenders, partially nested in the front fender, on each side. The frame was strengthened, and the wheelbase shortened compared to the civilian model; the independent suspension had helical springs in front and longitudinal torsion bars in the rear. The differential was lockable by the driver. The brakes were hydraulic, drum type on the four wheels. Lubrication and cooling systems were designed to allow the engine to operate in a high-temperature environment, and the carburettor air intake

The Alfa Romeo 2500 C used by Field Marshal Erwin Rommel. Unfortunately the images are of bad quality and some markings have been censored.

Alfa Romeo 2500 C and other vehicles in the Donetsk airfield, in 1942. On the left a S.81 Pipistrello bomber/cargo aircraft is visible.

The 2500 C sedan version on the Eastern Front.

was fitted with an oil-bath filter. The batteries for the electric starter were placed underneath the seats to protect them from heat. The 2500 C had four reserve fuel tanks; two of 25 litres each behind the rear fenders and two of ten litres each, triangular in shape, behind the spare wheels. The engine was a six-cylinder gasoline engine developing 87 HP. The manual transmission had four forward and one reverse gears, with third and fourth synchronised, and a locking differential.

Variants

An enclosed version (sedan) of the Alfa Romeo 2500 C was developed for use in Russia and the Balkans; 50 examples were built, using the chassis available for the torpedo version.

In 1943, Alfa Romeo proposed a 4x4 version of the 2500 C, with a reduction gear, but the proposal was not accepted.

Specifications

- Designation: Autovettura Alfa Romeo 2500 C
- Producer: Alfa Romeo, Milan
- Years produced: 1939–1945
- Number produced: 196
- Length: 4,700mm (15ft 5in)
- Width: 1,570mm (5ft 2in)
- Height: 1,600mm (5ft 3in)
- Unladen weight: 1,725kg (3,803lbs)
- Carrying capacity: Driver and four/six passengers
- Wheelbase: 3,100mm (10ft 2in)
- Front track: 1,445mm (4ft 8in)
- Rear track: 1,447mm (4ft 8in)
- Minimum turning radius: 5,750mm (18ft 10in)
- Minimum clearance: 260mm (10.23in)
- Tyres: Superflex 5.50 x 18
- Engine: Alfa Romeo, six-cylinder, water-cooled, 2,443cc, 54 HP @4,600 rpm
- Fuel: Gasoline
- Transmission: Four speeds forward, one reverse; 3rd and 4th gears synchronised
- Fuel capacity: 120 + 70 reserve in four supplementary tanks, 190 litres total (50 US gallons, 41.8 Imperial gallons)
- Drive layout: 4x2
- Maximum speed: 127km/h (79mph)
- Range (on road): 850km (528 miles) with supplemental fuel tanks

Autovettura Bianchi VM6 C

The series Bianchi VM6 C (Coloniale). (CSM/AUSSME)

Benito Mussolini and Marshal Pietro Badoglio at the Piccolo San Bernardo, a mountain pass in the Alps on the Italy-France border, on 28 June 1940. The VM6 C has Superflex Stella Bianca tyres. Note the circular badge indicating that the vehicle belongs to the *Regio Esercito* applied at the bottom of the windshield. (ACS)

A frame from a contemporary newsreel. The camouflage, visible on other military cars of the period, was probably in reddish-brown and dark green spots on a regulation grey-green base colour.

The same car as above. Note the dividing screen between the front and rear seats. (ACS)

Developmental and Service History

The Bianchi VM6 C (Coloniale) was derived in 1938 from the civilian Bianchi S6 sedan; it was built in only 100 copies. The VM6 C was intended for use as a staff car for high-ranking officers; it is known to have been in service in Russia, and probably in the Balkans, as well as in Italy itself. Some are known to have had machine guns mounted in the rear. Production ended in 1940.

Technical Description

The Bianchi VM6 C was a 4x2, rear-wheel drive open touring car. Its four half-doors had no glass windows, but there was a dividing screen between the front and rear seats whose ends could be rotated outwards. The squarish front windshield could be folded fully forward.

Compared to the civilian version, the military version had free-wheeling spares mounted behind the front mudguards. As happened in other cases, chassis and suspension had been reinforced. As befitted a vehicle that served as a command vehicle, the rear compartment had two small folding tables and three small storage compartments for maps and papers. Suspension was semi-elliptical springs and hydraulic shock absorbers; all four wheels had hydraulic brakes, and the hand brake acted on the rear

Another VM6 C showing the three-tone camouflage. (ACS)

wheels. Starting was electric. The engine was a six-cylinder gasoline engine developing 52 HP. The manual transmission had four forward and one reverse speeds.

Specifications
- Designation: Autovettura Bianchi VM6 C
- Producer: Edoardo Bianchi, Milan
- Years produced: 1938–1940
- Number produced: 100
- Length: 4,247mm (13ft 11in)
- Width: 1,680mm (5ft 6in)
- Height: 1,680mm (5ft 6in)
- Unladen weight: 1,500kg (3,307lbs)
- Carrying capacity: Four passengers
- Wheelbase: 2,785mm (9ft 2in)
- Front track: 1,420mm (4ft 8in)
- Rear track: 1,420mm (4ft 8in)
- Minimum turning radius: 5,120mm (16ft 8in)
- Minimum clearance: 230mm (9in)
- Tyres: Superflex 5.50 x 18
- Engine: Bianchi S9, six-cylinder, water-cooled, 2,179cc, 54 HP @4,000 rpm
- Fuel: Gasoline
- Transmission: Four speeds forward, one reverse
- Fuel capacity: 72 litres (19 US gallons, 15.8 Imperial gallons)
- Drive layout: 4x2
- Maximum speed: 102km/h (63mph)
- Range (on road): 380km (236 miles)

Autovettura Fiat 518 Coloniale

The Fiat 518 C Ardita series (short wheelbase). The body of this car and the other versions was designed by the coachbuilder Pininfarina.

The Fiat 518 L Ardita series (long wheelbase).

The Fiat 518 Coloniale (short wheelbase) also used by the *Regio Esercito*.

An Italian officer in playful pose on a 518 Coloniale.

Vans of the Reparti Fotocinematografici propaganda department following the troops. North Africa, 1941. (ACS)

Developmental and Service History

The Fiat 518 Coloniale civilian car was based on the Fiat 518 Ardita 2000 four-door convertible whose production began in 1933. The colonial version of the 518 was chosen by the army in anticipation of hostilities with Ethiopia to fill the need for a four-passenger automobile with a relatively powerful engine. Details concerning production numbers and service history are unavailable.

Technical Description

The Fiat 518 Coloniale had an open body with four doors; a canvas roof folded back for stowage behind the rear seats. The 518 was a standard configuration 4x2, with rear wheel drive; it was a left-hand driven vehicle. The 518 Coloniale was built both with a short wheelbase (*C Coloniale*, 2,700mm) and a long wheelbase (*L Coloniale*, 3,000mm) as well as the civilian versions 518 C (or *Corta*, short) 518 L (or *lunga*, long). Pneumatic tyres were mounted on pressed steel wheels; spare tyres mounted on wheels were secured on each side of the bonnet. Suspension consisted of solid axles, longitudinal leaf springs and hydraulic shock absorbers; all four wheels had hydraulic drum brakes. A battery and an electric starter were included. The engine was a four-cylinder gasoline engine developing 45 HP. The manual transmission had four forward and one reverse gears.

Variants

On the chassis of the Fiat 518 second series or the 527 Ardita – i.e. the 518 with an engine upgraded to 2,500cc – an unspecified number of vans were built for service with the *Reparti Fotocinematografici* propaganda department following the troops. The staff belonged to the Istituto Luce, an Italian film company founded in 1924 for the distribution of photographs and newsreels.

Specifications

- Designation: Autovettura Fiat 518 Coloniale
- Producer: Fabbrica Italiana Automobili Turin (Fiat), Turin
- Years produced: 1933–1938
- Number produced: Not Known
- Length: 4,241mm (13ft 3in) 518 C Coloniale; 4,542mm (14ft 11in) 518 L Coloniale:
- Width: 1,670mm (5ft 6in)
- Height: 1,670mm (5ft 6in)
- Unladen weight: 1,185kg (2,612lbs) 518 C Coloniale; 1,259kg (2,775lbs) 518 L Coloniale
- Carrying capacity: Four/six passengers
- Wheelbase: 2,700mm (8ft 10in)
- Front track: 1,180mm (3ft 10in)
- Rear track: 1,180mm (3ft 10in)
- Minimum clearance: 250mm (9.84in)
- Minimum turning radius: 5,800mm (19ft)
- Tyres: Superflex 5.50 x 20
- Engine: Fiat Type 118, four-cylinder, side-valve, water-cooled, 1,944cc, 45 HP @3,600 rpm
- Fuel: Gasoline
- Transmission: Four speeds forward, one reverse
- Fuel capacity: 62 litres (16.4 US gallons, 13.6 Imperial gallons)
- Drive layout: 4x2
- Speed: 82km/h (51 mph)

Autovetturetta Fiat 508 M

Developmental and Service History

In 1931 the Italian military authorities set out a request for a small model to be used as staff car and liaison vehicle. Based on this request, the next year Fiat offered a military version of a new model ready to be launched, the Fiat 508 compact car, colloquially known as the Balilla, in a two-door, two-seater version (aka Spider). Note that the civilian Balilla was built in many models over time: first series, the 508 A with three-speed transmission (later called the 508 A); second series, with a four-speed transmission (from 1934) called the 508 B; and a third and final series (the 508 C, from 1937) with a new body and engine which was later restyled and renamed the 1100 (1939).

The civilian first series Fiat 508 Balilla Torpedo (three-speed aka 508 A).

The civilian first series (three-speed) Fiat 508 A Spider. Note the two doors featured on this sporty model and the spare wheel fixed at the back.

A first-series 508 A Spider with *Regio Esercito* number plate in 1940. Compared to the previous photo, it has stamped disc wheels and the front bumpers are missing.

The 508 M two-door, two-seater, aka Spider Militare, military series version derived from the civilian 508 A Spider. Compared to this latter and the lack of bumpers, the larger and squared rear trunk, the stamped disc wheels with bigger tyres, and the spare wheel placed on the left side, are all clearly visible.

The civilian second series Fiat 508 Torpedo (four-speed, aka 508 B), recognisable by the slanted air intakes of the bonnet. This example mounts standard Pirelli Superflex Cord 4.00 x 17.

The 508 M, as the military version was designated, was acquired by the *Regio Esercito* in mid–1932, and the initial issue was to motorised division headquarters. The 508 M was first built on the basis of the 508 A (first series), then of the 508 B (second series). In addition to the *Spider Militare* (two-seat roadster, 1932), more different body styles were designed: *Torpedo Militare*, (four-seat touring, 1934), *Camioncino* (pick-up truck, 1933), and *Furgoncino* (small van, 1933).

The *Regio Esercito* acquired the Torpedo Militare and the Spider Militare to also assign them to the officers of the *Reali Carabinieri* (military police), from the rank of Major upwards if they did not have other service vehicles at their disposal. The greatest use of these cars was in the colonies of East Africa, where the distances to be covered were large.

Although production ceased in 1937, being replaced by the 508 CM based on the 508 C (third series) Balilla, many 508 M cars saw service in Italian East Africa and later in Spain, during the Civil War, and during the Second World War.

The 508 M and its variants was judged to be exceptionally stable, easy to drive and had good road handling qualities.

A 508 Torpedo Militare second series (four-speed). (D. Zambon)

The 508 M Spider second series, two-door, two-seater version. In both versions of the Spider Militare (three-speed and four-speed) a seat could be found in the trunk for a third passenger. The spare wheels were therefore placed at the body sides.

A camouflaged second series 508 M Spider.

Technical Description

The 508 M Spider Militare was a small two-seat open passenger car with a rear 'rumble seat' for a third passenger. It had a canvas top which could be deployed manually over the passenger compartment. Pneumatic tyres were mounted on pressed steel wheels; there were two spare wheels/tyres mounted on the sides of the bonnet. The windshield was of a one-piece design.

Compared to the civilian Spyder, the 508 M had larger diameter wheels and a modified transmission for greater traction in order to be able to climb steeper hills; offsetting that was the reduction in speed from 80–85km/h to 72–75km/h, and weight was slightly heavier.

The second series differed from the first in the shape of its radiator grille, the slanted rather than vertical bonnet cooling louvres, the wheelbase was 50mm longer and, importantly, a four-speed gearbox (with 3rd and 4th gears synchronised) replacing the three-speed gearbox.

From left to right: a 508 Furgoncino, a 508 Camioncino and a 508 Spider, all civilian versions first series (three-speed).

A 508 M Spider, probably in French territory in the summer of 1941.

This photograph of a 508 M Spider allows the fully raised top, including the sides, and the rear of the bodywork with the luggage trunk which, once opened, served as a folding seat, to be seen.

All four wheels had drum brakes. Suspension was leaf springs front and rear. The engine was a four-cylinder gasoline engine developing 20 HP in the 508 A and 24 HP in the 508 B.

Variants
The first series was followed by a second series (see above) with a slightly different body style and mechanical components.

The Torpedo Militare version launched in 1934 had four doors and could carry four – a driver and three passengers.

The 508 M Camioncino and Furgoncino variants are described in the chapter on Light Trucks, below.

The Coloniale version (i.e. with characteristics suitable for tropical and desert areas, at least on paper) of the different civilian Balillas was also used by the Italian armed forces and designated 508 M Col. or 508 M Coloniale.

Specifications
(Unless specifically noted, all measurements apply to 508 M aka Spider Militare version)
- Designation: Autovetturetta Fiat 508 M

A 508 Spider of the Royal Hungarian Army in 1939. Note the position of the passenger seated on the rear folding seat. (Fortepan 57463)

- Producer: Fabbrica Italiana Automobili Turin (Fiat), Turin
- Years produced: 1932–1939
- Number produced: Not Known
- Length: 3,250mm (10ft 8in)
- Width: 1,380mm (4ft 8in)
- Height: 1,450mm (4ft 9in) with top down; 1,600mm (5ft 3in) with top up
- Unladen weight: 690kg (1,521lbs)
- Carrying capacity: 240kg (529lbs)
- Wheelbase: 2,250mm (7ft 4in) 508 A; 2,300mm (7ft 6in) 508 B
- Front track: 1,180mm (3ft 10in)
- Rear track: 1,200mm (3ft 11in)
- Minimum turning radius: 4,750mm (15ft 7in)
- Minimum clearance: 220mm (9in)
- Tyres: Pirelli Superflex 4.00 x 17 or Michelin Stop 4.00 x 17 or Superconfort 400 x 135 on first series (508 A) and second series (508 B civilian); Superflex 4.40 x 19 on military 508 M; Aerflex 5.00 x 16 on 508 Coloniale
- Engine: Fiat 108 M four-cylinder, side-valve, water-cooled, 995cc, 20 HP @3,400 rpm 508 A; 24 HP @3,800 rpm 508 B
- Fuel: Gasoline
- Transmission: three speeds forward, one reverse 508 A; four speeds forward, one reverse (3rd and 4th gears synchronised) 508 B
- Fuel capacity: 26 litres (6.9 US gallons, 5.7 Imperial gallons)
- Drive layout: 4x2
- Speed: 75km/h (46.6mph)
- Range (on road): 290km (180) 508 A and 508 B; 310km (193 miles) 508 M

Autovettura Fiat 508 C and 508 C Coloniale

The Fiat 508 C (the third model of the civilian Balilla), Berlina or sedan version.

Two examples of civilian 508 C Balilla Berlina cars in service with the Italian forces in Libya in September 1940, and the Ukraine in August 1941. (ACS)

A Balilla Berlina with number plate of the *Regia Aeronautica* in 1941.

Balilla Berlina with number plate of the *Regio Esercito*.

Developmental and Service History

The Nuova Balilla Fiat 508 C (third series of the Balillas), civilian version, began to be produced in 1937 in several configurations – sedan, touring, convertible and cabriolet (by Viotti).

Two Balillas of different bodies in service with the Regio Esercito. Note the different shape of the number plates.

The Fiat 508 C Coloniale, a four-door convertible intended for use in overseas colonies.

In 1939 the 508 C underwent a restyling, mainly in the grille and in the bonnet; from then on its name became the *Millecento* (one thousand and one hundred) or 1100 in reference to its 1,089 (1,100) cubic centimetre engine. This model was also produced with limited modifications, under licence in France by Simca and in Germany by NSU.

The redesigned version 1100, because of the bulky front, was nicknamed *Musone* (big snout). So, the Nuova Balilla 508 C early model was afterwards christened *Musetto* (small snout).

Initially, the Italian armed forces incorporated both the 508 C Berlina, i.e. the sedan civilian version with four seats and four doors, and its colonial variant, a convertible model slightly modified for the use in the African colonies. The version for military purposes was based on the colonial civilian variant, with a few modifications, and for this reason it was christened 508 C Militare Coloniale (in short, 508 Mil. Col.). It featured two spare wheels at the rear. This car finally represented a transitional model and only a few examples were produced for the *Regio Esercito*. But their positive features, namely the low fuel consumption rate, high stability and good speed, despite traction only on the rear wheels, prompted the military authorities to ask Fiat for a new torpedo all-terrain version.

So, in mid–1938 the definitive model of the military car was produced, which was renamed the 508 C Torpedo Militare or, more briefly, the 508 CM (i.e. third series, military). This is described below.

Technical Description

The civilian Nuova Balilla, initialled 508 C, featured a completely different bodywork from those of the two previous series (508 A and 508 B, as they were called from that moment on). Even the engine, the Fiat 108C, was redesigned and was more powerful. The gearbox had four gears, plus reverse, and the car could reach 110km/h. Another difference compared to the previous series

Two views of a Fiat 508 C Militare Coloniale. It was almost identical to the civilian coloniale, except for the two rear spare wheels.

The Fiat 1100 Musone, i.e. the restyled 508 C civilian sedan.

lay in the suspension: the front suspension had independent wheels equipped with coil springs assisted by hydraulic shock absorbers; the rear ones remained of the leaf spring type. The 1100, apart from the bodywork, was basically identical.

Variants
A long wheelbase version, called 508 L and capable of carrying six passengers, was also manufactured.

Specifications
- Designation: Autovetturetta Fiat 508 C
- Producer: Fabbrica Italiana Automobili Turin (Fiat), Turin
- Years produced: 1938–1953
- Number produced: more than 57,000 examples

- Length: 3,615mm (11ft 10in)
- Width: 1,480mm (4ft 10in)
- Height: 1,480mm (4ft 10in); 1630mm (5ft 4in) with top closed
- Total weight: 1,160kg (2,557lbs)
- Carrying capacity: 4 passengers – 300kg (661lbs)
- Wheelbase: 2,420mm (7ft 11in)
- Front track: 1,231mm (4ft ½in)
- Rear track: 1,226mm (4ft)
- Minimum turning radius: 4,500mm (14ft 9in)
- Minimum clearance: 170mm (6.7in)
- Tyres: Superflex 5.00 x 15 Berlina; 5.00 x 17 Coloniale
- Engine: Fiat 108 C, four-cylinder, valve-in-head, water-cooled, 1,089cc, 32 HP @4,400 rpm
- Fuel: Gasoline
- Transmission: Four speeds forward, one reverse
- Fuel capacity: 32 litres (8.45 US gallons, 7 Imperial gallons)
- Drive layout: 4x2
- Speed: 110km/h (68mph)
- Range (on road): 370km (230 miles)

Autovettura Fiat 508 CM (1100 Mimetica)

Note

To avoid confusion with the models described above, in this entry the denominations 508 C Torpedo Militare and 508 C Torpedo Coloniale have been used, instead of 508 Militare and 508 C Militare Coloniale, although the latter two are also present in the documentation issued by Fiat.

The prototype of the Fiat 508 CM. (Fiat)

Developmental and Service History

As previously mentioned, towards the end of the thirties the Italian military authorities asked Fiat to design a new off-road car in a torpedo configuration more suitable for use in war zones than the 508 M, which had nonetheless impressed.

The Turin company started from the chassis of the new Balilla, i.e. the third series or Fiat 508 C (see above). In mid-1938, the civilian car body was completely redesigned to meet military specifications and requirements thus taking on a much more 'martial appearance'. This became the definitive version of the military car, which was renamed the 508 C Torpedo Militare or, more briefly, the 508 CM (i.e. third series, military). It had a sturdier chassis, reinforced suspension, increased ground clearance and modified transmission with a top speed limited to 95km/h. The wheels were also larger. But the most distinctive element was the squared-off body. The early definitive 508 CM had a camouflage paint scheme, and – considering the change of name of the civilian version – was thus often referred to as Fiat 1100 Mimetica (camouflaged).

Fiat 508 CM (aka Fiat 1100 Mimetica) staff car, Italy, 1939. (Artwork by and © David Bocquelet)

Autocarretta 35 light truck, mountain chain of the Alps, Italy, 1935. (Artwork by and © David Bocquelet)

Fiat 618 light truck, East Africa, 1936. (Artwork by and © David Bocquelet)

SPA 38R light truck, Eastern Front, 1941. (Artwork by and © David Bocquelet)

COLOUR SECTION III

SPA CL 39 light truck, Eastern Front, 1941. (Artwork by and © David Bocquelet)

SPA AS 37 (second series) light truck, North Africa, 1942. (Artwork by and © David Bocquelet)

Benito Mussolini and Marshal Pietro Badoglio on board a Bianchi VM6 C staff car in a newsreel frame from 28 June 1940.

Photograph taken by a German soldier during the transfer trip and stay in North Africa. On the left, an Opel Olympia 38 that appears camouflaged in the *Tropen* colours employed by the Wehrmacht since March 1942. On the right, a Fiat 508 C Camioncino belonging to the *Regio Esercito* whose grey-green colour is only partially covered by the *kaki sahariano* (sand-yellow) applied by spray gun.

A 508 L Camioncino, painted grey-green.

An advertising poster for the Breda company (1931).

The cover of a 1934 motorcycling magazine with Sertum company propaganda.

The updated version of the Viberti brand.

The original brand of coachbuilder Viberti.

The Fiat 508 CM, aka 508 C Torpedo Militare, aka Fiat 1100 Mimetica. (Fiat)

A 508 CM on the assembly line in 1940. (LIFE)

The 508 CM / 1100 Mimetica was produced from 1938 to 1945 and saw service on all fronts in which the Italians were engaged. It was generally issued to the motor pools of headquarters elements and was extensively used by officers. It was the most widely used liaison and reconnaissance vehicle in the *Regio Esercito* and the *Regia Aeronautica*. About 6,000 examples of the 508 CM were built until 1943, and a further 1,354 examples were delivered to the German forces after the Armistice. The surviving vehicles served in Italy until 1960.

Technical Description

The 508 CM, aka 1100 Mimetica, was a four-door, four-seater, 4x2, right-hand, rear-wheel drive car. It was normally seen used as an open car, but it did have a retractable canvas roof and side curtains, with clear window inserts made of celluloid. The windshield was of a one-piece type with a single electric windshield wiper for the driver. The wheels were pressed steel discs with

Factory photo of the Fiat 508 CM or 1100 Mimetica. (Fiat)

An artillery officer aboard a Fiat 508 CM aka 1100 Mimetica.

Mussolini reviews the troops aboard a Fiat 1100 Mimetica.

Fiat 1100 Mimetica. (D. Zambon)

Bir El Gubi, Libya, autumn 1941. A Fiat 508 CMC, aka Fiat 1100 Torpedo Coloniale, together with motorcycles, trucks and AB 41 armoured cars. The 'balloon' tyres and the yellow sand colour confirm that this is a colonial variant, (ACS)

A small 4x2 car like the 508 CM was unable to cope with the mud of the Eastern Front. (ACS)

small hub caps; a spare wheel and tyre were bolted onto the rear of the body. The chassis consisted of two longerons of pressed sheet metal; it was narrower in the front and was made rigid by a double cross-member in the shape of an 'X'. The front bumper was attached to the longerons. The independent front suspension consisted of helical springs with hydraulic shock absorbers; rear suspension consisted of a solid rear axle, leaf springs, hydraulic shocks and a stabiliser bar. Brakes were four-wheel hydraulic drums and a hand brake acting on the transmission. The electrical system was a 12 volt system with battery and electric starter. The engine was a four-cylinder gasoline engine developing 30 HP. The manual transmission had four forward and one reverse gears.

Variants

The 508 CM conceived for the North African theatre, named 1100 Torpedo Coloniale or 508 CMC, manufactured until 1943, was fitted with special filters for the desert environment, larger fuel tank (70 litres instead of 40 litres), enhanced battery, modified transmission, different wheel track and ground clearance. The standard, low pressure Superflex Cord 5.00 x 18 pneumatic tyres were replaced with very low pressure Aerflex 6.00 x 16 balloon tyres, smaller but wider and thus more suitable for soft and sandy terrain. Except for these elements and the colour, which was the *kaki sahariano* (sand-yellow) specific to that theatre of war, it was almost identical to the standard 1100 Mimetica.

Siliana, Tunisia, April 1943. A posed photo of French Goumiers of the 2eme Tabor of the 1st GTM (Group of Moroccan Tabors) of the 1st DMM (Moroccan Marching Division) aboard a captured FIAT 508 CM. (ECPAD)

The Fiat 508 CM Berlina FO (*Fronte Orientale*, i.e. Eastern Front) was a completely enclosed and heated sedan, resembling the civilian model. 50 examples were assigned to the Italian expeditionary corps in Russia in 1941.

The 508 CM frame was used as the basis for the Fiat 1100 Camioncino pick-up truck (see below).

Specifications

- Designation: Fiat 508 Torpedo Militare or Fiat 508 CM
- Producer: Fabbrica Italiana Automobili Turin (Fiat), Turin
- Years produced: 1938–1945
- Number produced: more than 7,000 examples
- Length: 3,615mm (11ft 10in)

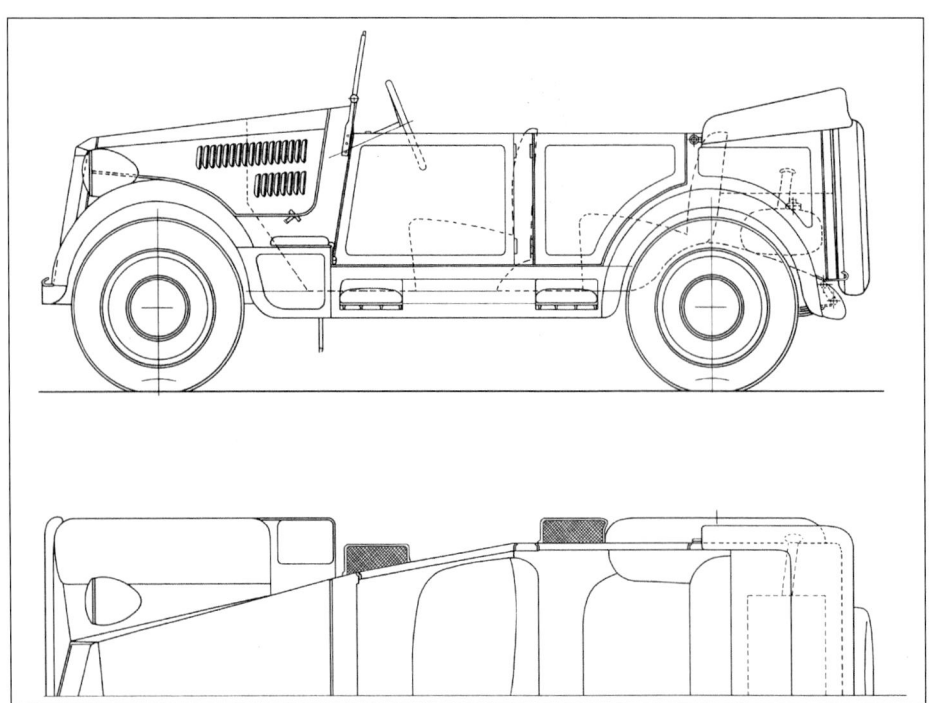
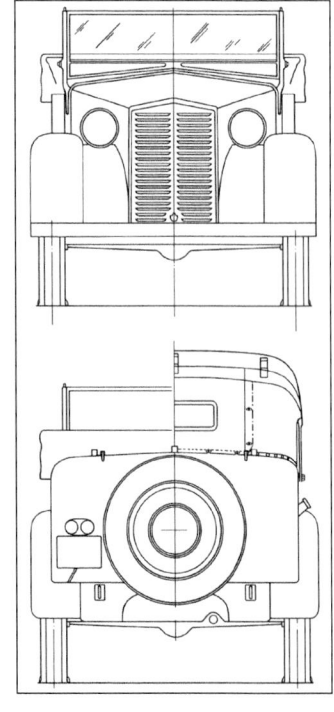

Fiat 508 CM. (Drawings by GMT)

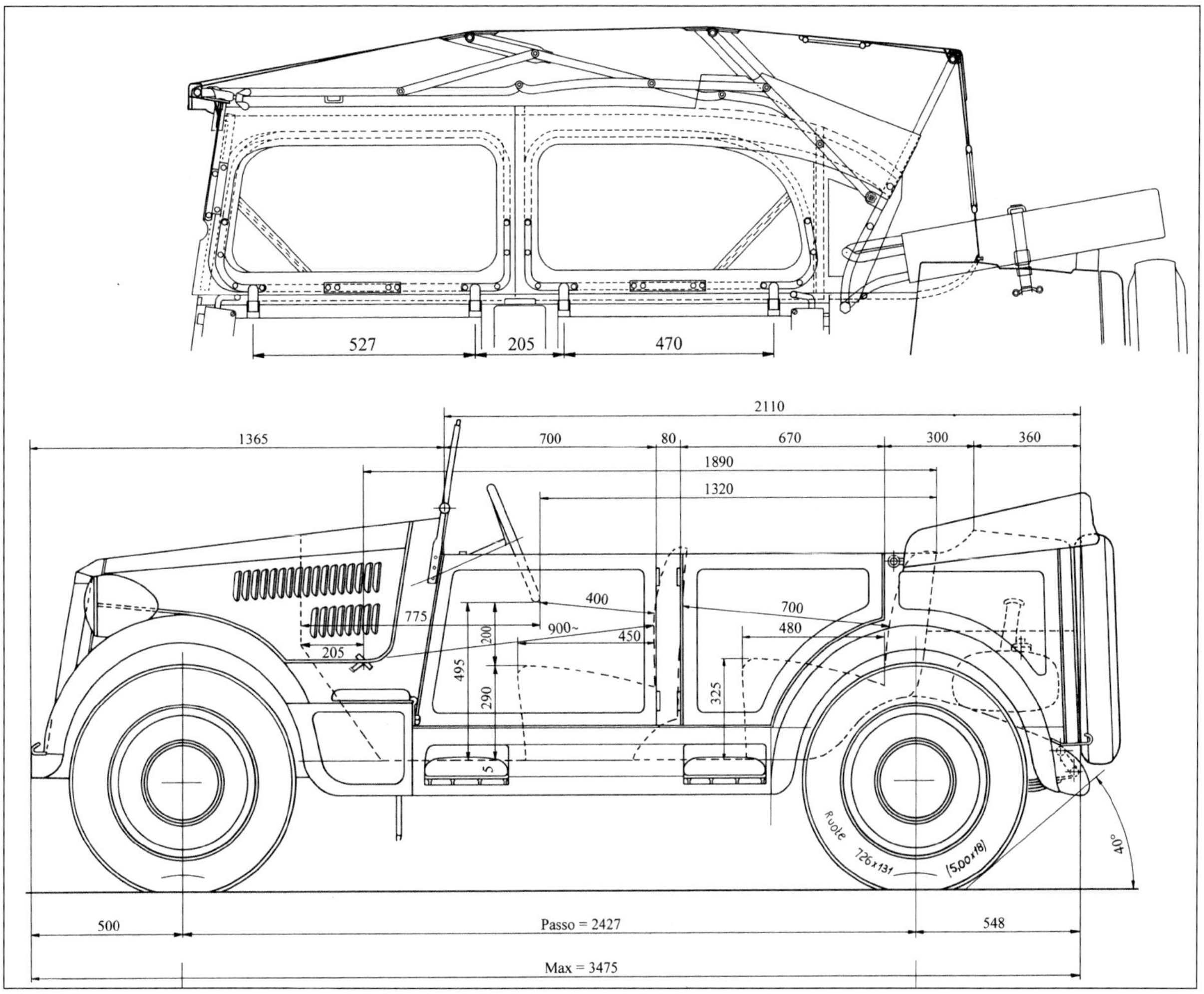

Original Fiat drawings with dimensions of the car and the retractable canvas roof curtain (left side).

Fig. 1. — Autotelaio Mod. 508 C Militare.

Transparent view taken from the user manual.

We do not know if this Fiat 508 CM Coloniale with special equipment, conceived for special units similar the British Long Range Desert Group (LRDG), was ever employed or was just a proposal. It carries pioneer tools, additional canisters, a searchlight on the left side but lacks a windshield. Additionally, it is armed with a 6.5mm Breda 30 light machine gun. Inside, the barrels of some other weapons are visible, including that of a MAB 38 A submachine gun. (C. Pergher)

- Width: 1,480mm (4ft 10in)
- Height: 1,480mm (4ft 10in); 1630mm (5ft 4in) with top closed
- Unladen weight: 890kg (1,962lbs)
- Carrying capacity: 300kg (661lbs)
- Wheelbase: 2,427mm (7ft 11in)
- Front track: 1,242mm (4ft) Torpedo Militare; 1,256mm (4ft 2in) Torpedo Coloniale
- Rear track: 1,239mm (4ft) Torpedo Militare; 1,253mm (4ft 1in) Torpedo Coloniale
- Minimum turning radius: 5,000mm (16ft 5in)
- Minimum clearance: 237mm (9¼in) Torpedo Militare; 230mm (9in) Torpedo Coloniale
- Tyres: Superflex 5.00 x 18 Torpedo Militare; Aerflex 6.00 x 16 Torpedo Coloniale
- Engine: Fiat 108 C, four-cylinder, valve-in-head, water-cooled, 1,089cc, 30 HP @4,400 rpm
- Fuel: Gasoline
- Transmission: Four speeds forward, one reverse
- Fuel capacity: 40 litres (10.57 US gallons, 8.8 Imperial gallons) Torpedo Militare; 70 litres (18.5 US gallons, 15.4 Imperial gallons) Torpedo Coloniale
- Drive layout: 4x2
- Speed: 95km/h (59mph)
- Range (on road): 320km (201 miles) Torpedo Militare; 570km (354 miles) Torpedo Coloniale

Autovetturetta Fiat 500 Topolino

Developmental and Service History

Written and photographic documentation tell us that the small Fiat 500 car was never officially adopted by the Italian armed forces. However, some examples may have been purchased by the Italian Army, Air Force or Navy for assignment to officers or for liaison duties in metropolitan areas. Although, probably, other cars – such as the Fiat 508 in all its variants – were preferred.

The 500 project started in 1934; the Turin-based manufacturer felt confident from its experiences with the 509 and 508 models and decided to create 'a small two-seater capable of making motoring more popular in Italy', as the owner Giovanni Agnelli declared. The story that it was Mussolini who personally requested such a small and cheap car is unconfirmed, but it nevertheless falls within the climate of that period.

The Fiat 500, launched in 1936, was later nicknamed *Topolino*, because of its minute and tapered shape – the Italian translation of 'Mickey Mouse' (*Topolino*) actually had nothing to do with the car, although it later helped to spread the name more widely.

As happened with almost all the Italian civilian and military vehicles, a great number of examples were requisitioned by the Germans after the Armistice. This model is also called Fiat 500 A to differentiate it from the 500 B and 500 C models produced after the end of the war.

The Fiat 500 was also produced in France by Simca (with the name Simca 5), in Poland by Polski-Fiat and in Germany by NSU, all with small external and internal differences in comparison to the original Italian model.

A civilian Fiat 500 Topolino in front of the Ente Italiano per le Audizioni Radiofoniche, the Italian National Radio Broadcasting Station in Asiago Street, during the operations in Rome following the Armistice of 8 September 1943.

A Fiat 1100 pick-up and a Fiat 500 Topolino Trasformabile (with folding top) employed by 3. Batterie, I. Abteilung, Artillerie-Regiment Hermann Göring, LW Panzer-Division Hermann Göring.

A 500 Topolino used by Stug-Kompanie of the Pz.Jg.Abt. 46, Reichsgrenadier-Division Hoch und Deutschmeister in late 1943. The vehicle received new markings, number plate and camouflage from the new owners.

A Fiat 500 Topolino from the German 26. Panzer-Division in Formignana, near Ferrara, in January 1945. Note the 'FIAT' logo on the wheel rim cover. It is an example produced after 1938, with reinforced rear suspension, turn indicators on the windscreen struts and the interior rear mirror at the top. Some details, such as the external mirror and the two-piece bumpers – not included in the standard models in Italy, see above – suggest a special version or one produced for the foreign market.

A huge German Büssing-NAG Typ 650 together with a Fiat 500 A Topolino car. Both vehicles belonged to a headquarters and shown a complete multi-coloured camouflage. The car had an improvised, but official, German number plate.

Milan, a Repubblica Sociale Italiana column – which includes a Bianchi Miles truck and an M40 tank – at the halt among Italian and German soldiers. The Fiat 500 Topolino was a common sight during the Italian Campaign. (S. Di Giusto)

The Fiat 500 short leaf spring chassis furgoncino (small van). (Fiat)

Technical Description

The first model produced, the Fiat 500, could reach a speed of 85km/h and carry two adults and a 50kg bag or two children. It weighed 535kg and, thanks to a 569cc engine, consumed 6 litres of gasoline per 100km. The chassis consisted of two longerons of pressed steel plate, with lightening holes, connected by two cross-members. The front wheels had independent suspension with upper transverse leaf spring and hydraulic shock absorbers. The suspension of the rear wheels were half-spring in the first series (*balestra corta*, short spring), and full leaf springs in the second series (*balestra lunga*, long spring) with hydraulic shock absorbers.

Variants

The Fiat 500 Trasformabile (convertible) had a folding top.

A van with a capacity of 300kg (Furgoncino) was built on the chassis of the Fiat 500. The version on short leaf spring chassis had only one inclined rear door, the version on the long leaf spring chassis, reinforced compared to the previous one, had two vertical rear doors.

A Fiat 500 long leaf spring furgoncino of the Fallschirm-Sanitäts-Abteilung 4 from 4. Fallschirmjäger-Division, a unit formed in Italy in November 1943. The white rampant horse within a black circle is visible on many Italian vehicles used by the German Medical Corps, which were repainted completely white.

The Fiat 500 A Topolino. (Drawings by A. M. Feller – GMT)

Specifications

- Designation: Autovetturetta Fiat mod. 500
- Producer: Fabbrica Italiana Automobili Turin (Fiat), Turin
- Years produced: 1936–1948
- Number produced: 122,213
- Length: 3,215mm (10ft 6in)
- Width: 1,275mm (4ft 2in)
- Height: 1,377mm (4ft 6in)
- Unladen weight: 535kg (1,180lbs)
- Wheelbase: 2,000mm (6ft 7in)

- Front track: 1,114mm (3ft 8in)
- Rear track: 1,083mm (3ft 6in)
- Tyres: Superflex 4.00 x 15
- Engine: Fiat 500 four-cylinder, side-valve, water-cooled, 569cc, 13 HP @4,000 rpm
- Fuel: Gasoline
- Transmission: four speeds forward, one reverse (3rd and 4th gears synchronised)
- Fuel capacity: 21 litres (5.5 US gallons, 4.6 Imperial gallons)
- Drive layout: 4x2
- Speed: 85km/h (53mph)

Autovettura Fiat 2800 CMC

Benito Mussolini greets King Vittorio Emanuele III aboard his Fiat 2800 Torpedo Reale. (ACS)

The prototype of the Fiat 2800 Corta Militare Coloniale. (Fiat)

The series Fiat 2800 CMC during tests in Libya in 1939. Compared with the prototype, note the spare wheels positioned at the sides, and the differences in the body and retractable curtain. (Fiat)

Developmental and Service History

The Fiat 2800 CMC (the CMC stood for *Corta Militare Coloniale*, or short military colonial) was a military version of the somewhat elegant 2800 four-door touring car produced from 1938 to 1944. The civilian Berlina and Torpedo versions of the car were widely used as official vehicles, and among other notables, were used to carry King Vittorio Emanuele III, Prince Umberto of Savoia, Mussolini and the Pope. Some 624 2800s chassis were built, 210 of which were for Italian military use and intended for high-ranking officers. A few survived the war and served in the post-war Italian Army until replaced by more modern vehicles.

Technical Description

The Fiat 2800 CMC was smaller and incorporated a number of improvements and features not found in the civilian version of the 2800. The 2800 CMC was a four-door, rear-wheel drive touring car (with a body similar to the 508 CM body) with a retractable canvas top. The all-metal body had prominent curved fenders; an early version carried two spare tyres on the rear of the body, while the definitive version moved the two spares to the sides of the bonnet, directly behind and partially sunk into the front fenders. Mechanically, the 2800 CMC had a reinforced shorter frame (3.0 metres instead of 3.2 metres), heightened ground clearance, was fitted with an air filter, oil filters, an auxiliary electric fuel pump, two 6 volt batteries, a modified transmission, bigger tyres, modified hand brake and a transversally mounted muffler; it also had a 130 litre fuel tank. The traction was rear, with four-speed manual gearbox and hydraulic brakes. The independent front suspension consisted of helical springs and hydraulic shock absorbers; rear suspension was a solid axle with longitudinal leaf springs and hydraulic shock absorbers. The engine was a six-cylinder gasoline engine developing 85–90 HP. The manual transmission had four forward and one reverse gears.

Another image of the Fiat 2800 CMC during tests. It carries an unusual civilian licence plate of the Government of Libya. (Fiat)

The 2800 CMC with Prince Umberto on board passes among the troops of the 101ª *Divisione Trieste* on the Alps, in June 1940, during the attack on France.

Variants
At the beginning of 1943 the prospect of a variant of the 2800 CMC with four-wheel drive was considered, but the project was never completed.

Specifications
- Designation: Autovettura Fiat 2800 CMC
- Producer: Fabbrica Italiana Automobili Turin (Fiat), Turin
- Years produced: 1939–1944
- Number produced: 624 chassis (about one-third for the CMC)
- Length: 4,795mm (15ft 9in)
- Width: 1,884mm (6ft 2in)
- Height: 1,377mm (4ft 6in); 1,768mm (5ft 10in) with top closed
- Unladen weight: 1,970kg (4,343lbs)
- Carrying capacity: four/six (driver and three passengers, plus two additional jump seats for an additional two passengers)
- Wheelbase: 3,000mm (9ft 10in)
- Front track: 1,452mm (4ft 9in)
- Rear track: 1,460mm (4ft 9in)
- Minimum turning radius: 6,100mm (20ft)
- Minimum clearance: 245mm (10in)
- Tyres: Superflex 4.00 x 18 or 6.50 x 18
- Engine: Fiat 2852 MC six-cylinder, water-cooled, 2,852cc, 90 HP @4,400 rpm
- Fuel: Gasoline
- Transmission: Four speeds forward, one reverse, manual transmission
- Fuel capacity: 130 litres (34 US gallons, 28.6 Imperial gallons)
- Drive layout: 4x2
- Speed: 115km/h (71mph)
- Range (on road): 400km (249 miles)

AUTOVETTURA LANCIA APRILIA COLONIALE

Developmental and Service History
On the eve of the war, the *Regio Esercito* was desperately short of motor vehicles of all types and classes. Lancia offered a militarised version of its Lancia Aprilia sedan car in 1941 and conceived it as being for officers' use. The prototype underwent several modifications, leading to a somewhat simplified definitive version. 251 examples were built between 1941 and 1943.

The Aprilia Coloniale was mainly used by general officers. In service, the car suffered frequent problems with its suspension, steering gear, clutch and radiator. The Aprilia Coloniale saw service on all fronts, and following the 8 September 1943 Armistice, was used to some extent by German forces in Italy.

Technical Description
The Lancia Aprilia Coloniale was a torpedo (touring) car based on the civilian Lancia Aprilia Trasformabile second series, on a modified Lancia 639 S frame, shortened and reinforced in comparison to the civilian model. The body was designed and built

A first series Lancia Aprilia sedan with Fergat 'Littoria' spoked wheels and Michelin tyres.

A Lancia Aprilia Convertible with a Polish licence plate in 1952. (Maciek Peda)

The prototype of the Aprilia Coloniale. (Lancia)

by Viotti, a well-respected coach builder based in Turin. The body itself was a four-door touring type with a retractable canvas top; the window openings were fitted with side curtains and clear window inserts. A feature of the Aprilia Coloniale was a rifle rack behind the driver's seat which could hold four mod. 91 carbines.

The Aprilia Coloniale was a 4x2 vehicle with rear-wheel drive and driven right-hand. Pneumatic tyres were mounted on pressed 'Littoria' type steel disc wheels; a spare wheel and tyre were carried on the rear of the body. The prototype had removable auxiliary fuel tanks, but the production version had 20-litre fuel canisters carried behind the front and rear fenders on each side.

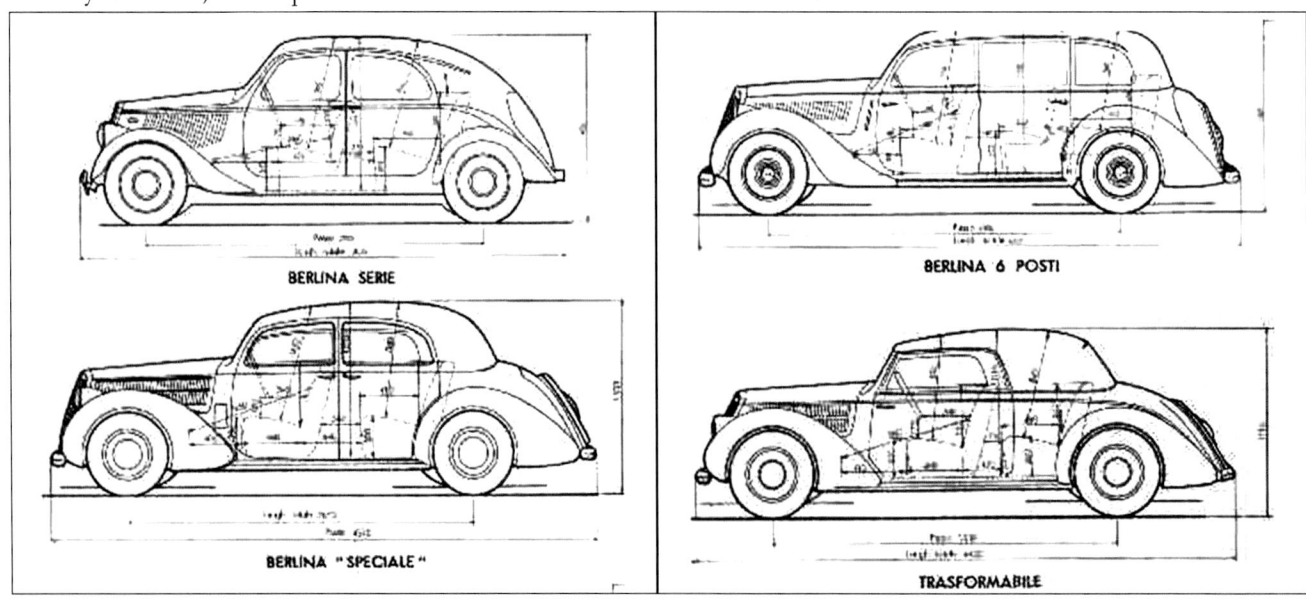

Image taken from a Lancia catalogue which schematically illustrates the four versions of the first series.

Factory pictures of the Lancia Aprilia Coloniale. Note the four 20-litre jerry cans located on the sides, front and rear. The bodywork was very similar to that of the Fiat 508 CM, which sometimes makes it difficult to distinguish the two vehicles in photographs. (Lancia)

The independent front suspension consisted of helical springs and hydraulic shock absorbers; rear suspension was a solid axle with longitudinal leaf springs and hydraulic shock absorbers. The electrical system was a 6-volt system with battery. The engine was a four-cylinder gasoline engine developing 51 HP. The manual transmission had four forward and one reverse gears; the clutch was a single plate dry disc.

The internal rack for rifles and the reclining rear seat. (Lancia)

A Lancia Aprilia Coloniale leading a vehicle convoy crossing a bridge built by Italian engineers in the Ukraine. (ACS)

General Alessandro Gloria, commander of the Divisione Bologna. (G. Forbicini)

Specifications
- Designation: Autovettura Lancia Aprilia Coloniale
- Producer: Lancia & C. Fabbrica Automobili, Turin
- Years produced: 1941–1943
- Number produced: 251
- Length: 4,326mm (14ft 2in)
- Width: 1,550mm (5ft)
- Height: 1,650mm (5ft 5in)
- Unladen weight: 1,112kg (2,452lbs)
- Carrying capacity: Four passengers, including the driver
- Wheelbase: 2,650mm (8ft 8in)
- Front track: 1,233mm (4ft)
- Rear track: 1,292mm (4ft 3in)
- Minimum turning radius: 5,100mm (16ft 9in)
- Minimum clearance: 226mm (9in)
- Tyres: Superflex 13 x 45
- Engine: Lancia Tipo 99 four-cylinder V, water-cooled, 1,485cc, 46 HP @4,000 rpm
- Fuel: Gasoline
- Transmission: Four speeds forward, one reverse
- Fuel capacity: 75 litres, plus four 20-litres fuel canisters (80 litres), total 155 litres (41 US gallons, 34 Imperial gallons)
- Drive layout: 4x2
- Maximum speed: 110km/h (68.3 mph)
- Range (on road): 620km (385 miles) with supplemental fuel canisters

Autovettura Lancia Artena Militare

The civilian Lancia Artena fourth series Ministeriale (#341 chassis) by Pininfarina.

The Lancia Artena Militare 6 luci 4 porte by Viotti.

The Lancia Artena Militare Trasformabile, aka 4 luci 4 porte.

Developmental and Service History

The civilian Lancia Artena was manufactured from 1931 to 1936 in three different production series. The fabrication of a military version, based on the Lancia 341 frame (fourth series), was resumed in 1940 for the *Regio Esercito* to make up for the shortage of military vehicles. Between 1940 and 1942 a total of 507 were built. The car was assigned to high-ranking officers at army group, army and corps level headquarters. Mussolini travelled in an Artena during a visit to the Greek-Albanian Front in 1941.

The Lancia Artena Militare Torpedo Trasformabile with grey-green or camouflage colouring. Note the partially foldable hood.

Technical Description

The Lancia Artena was built on a Lancia 341 frame (fourth series), and included three six-seat body styles:

1) the Berlina, which was a four-door six-window saloon, referred to as the Ministeriale (civilian variant) or the 6 luci 4 porte (six-windows four-door, military variant);

2) the Trasformabile 4 luci 4 porte, a four-window four-door touring car with a folding roof;

3) the Torpedo Trasformabile which was a four-door touring with a more squarish and simplified body, intended for the field use.

Viotti of Turin, which had close ties to Lancia, was commissioned to build the bodies of the military version, while Pininfarina took charge of designing the *Ministeriale*.

The Lancia Artena Militare was a 4x2 vehicle with rear-wheel drive and right-hand drive. Pneumatic tyres were mounted on 'Littoria' style disc wheels made by Fergat, a specialised company based in Turin. A spare wheel and tyre were mounted directly behind the front fenders on each side. The front suspension was independent with hydraulic shock absorbers; rear suspension was a solid axle with semi-elliptical leaf springs and mechanical shock absorbers. All wheels had hydraulic drum brakes. The engine was a water-cooled four-cylinder gasoline engine developing 51 HP. The manual transmission had four forward and one reverse gears; the clutch was a single plate dry clutch. The electrical system used a 12-volt battery.

Variants
As described above. Additionally, 191 were reportedly built with an ambulance body made by Bertone on a different frame – the model 441.

Specifications
- Designation: Autovettura Lancia Artena 4a Serie – Tipo Militare
- Producer: Lancia & C. Fabbrica Automobili, Turin
- Years produced: 1940–1942
- Number produced: 316 (plus 191 with ambulance frame)
- Length: 4,960mm (16ft 3in)
- Width: 1,730mm (5ft 8in)
- Height: 1,750mm (5ft 9in)
- Unladen weight: 956kg (2,107lbs)
- Carrying capacity: six passengers (two of whom were on front facing folding seats in the Trasformabile and Torpedo Trasformabile versions)
- Wheelbase: 3,180mm (10ft 5in)
- Front track: 1,400mm (4ft 7in)
- Rear track: 1,420mm (4ft 8in)
- Minimum turning radius: 5,170mm (17ft 2½in)
- Minimum clearance: 200mm (8in)
- Tyres: Superflex 15 x 45 or 16 x 45
- Engine: Lancia Tipo 84, four-cylinder V, water-cooled, 1,924cc, 51 HP @3,800 rpm
- Fuel: Gasoline
- Transmission: Four speeds forward, one reverse, manual transmission
- Fuel capacity: 70 litres (18.5 US gallons, 15.4 Imperial gallons)
- Drive layout: 4x2
- Maximum speed: 105km/h (65 mph)
- Range (on road): 380km (236 miles)

The Italian armed forces, in particular the *Regia Aeronautica* and, to a lesser extent, the *Regio Esercito* and the *Regia Marina*, employed a number of civilian executive cars such as the Fiat 1500, a model produced from 1935.

Fiat 1500 cars from *Regia Aeronautica* and *Regia Marina*. The three standing officers are Spanish pilots of the Ejército del Aire photographed in Caserta in 1939. The Royal Navy car, with 'RM 2467' number plate and special Pininfarina bodywork, was photographed in La Spezia before the war.

4

Light trucks

Autocarretta OM 32, 35 and 36

The Autocarretta 32, right-hand side view. (CTM)

The Autocarretta 35, left-hand side view. (CTM)

Developmental and Service History

The *autocarretta* (motorised cart) was a uniquely Italian light vehicle, designed expressly for employment in Italy's mountainous terrain, but later used in other roles as well. In 1927 the Italian military authorities set out a request for a vehicle that would be able to operate on bad roads and rough terrain. In fact, one of the official descriptions would later define it as a 'mountain reconnaissance car'.

Prototypes presented by Pavesi and Alfa Romeo were deemed unsuitable, followed by projects from Fiat, Lancia, Ansaldo and Ceirano. The Ansaldo entry was accepted, and a prototype was tested at the end of 1929. However, the *Società Anonima*

LIGHT TRUCKS 99

Rear view of the Autocarretta 32 (left) and 35 (right) compared. The latter shows a lower height achieved from the wider wheel track and modified suspension. (CTM)

The Autocarretta 35, front view. The central groove of the semi-pneumatic tyres favoured the fastening of the devices to improve grip on the ground. (CTM)

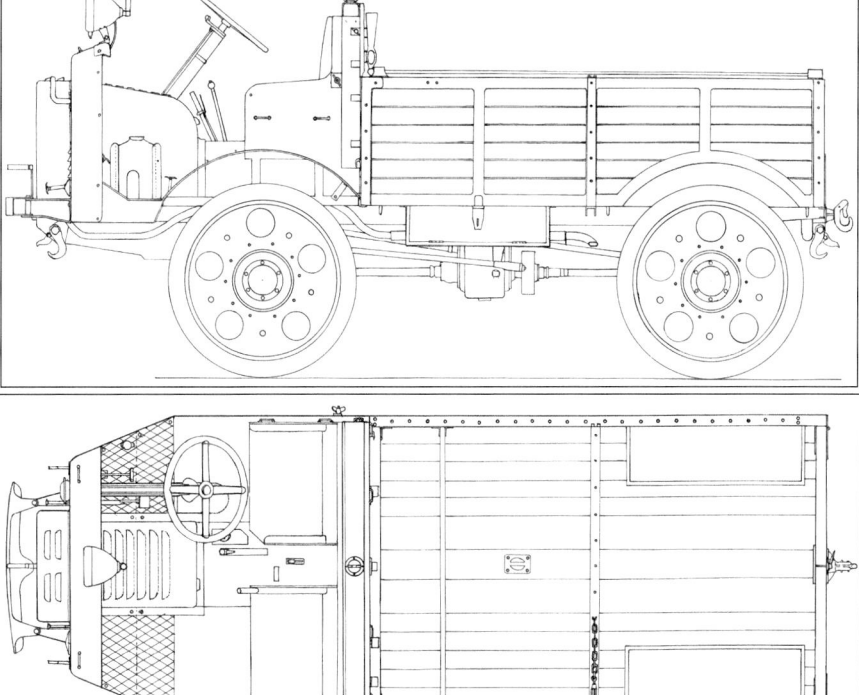

The Autocarretta 35. (Drawings by N. Pignato via GMT)

Automobili Ansaldo, based in Turin (not to be confused with the Genoese industrial group of the same name), was in dire straits and was taken over by the Officine Meccaniche (OM) of Brescia, which also assumed the manufacturing contract for the small vehicle. In the second half of 1931, OM furnished three test vehicles, slightly different from the Ansaldo prototype, which were subjected to severe tests in different conditions, obtaining homologation as the Autocarretta 32 at the beginning of the following year. A batch of pre-production models of these vehicles was issued to units for field testing; some of them took part in the large-scale manoeuvres of August 1932 with great success.

In mid-1932, a second batch of these vehicles in an updated design was ordered, with deliveries completed in 1935. From the applied modifications, dictated by previous experiences, a new model emerged, designed the Autocarretta 35 (or second series). The improvements made the vehicle more stable, thanks to better suspensions and a wider wheel track, and with better

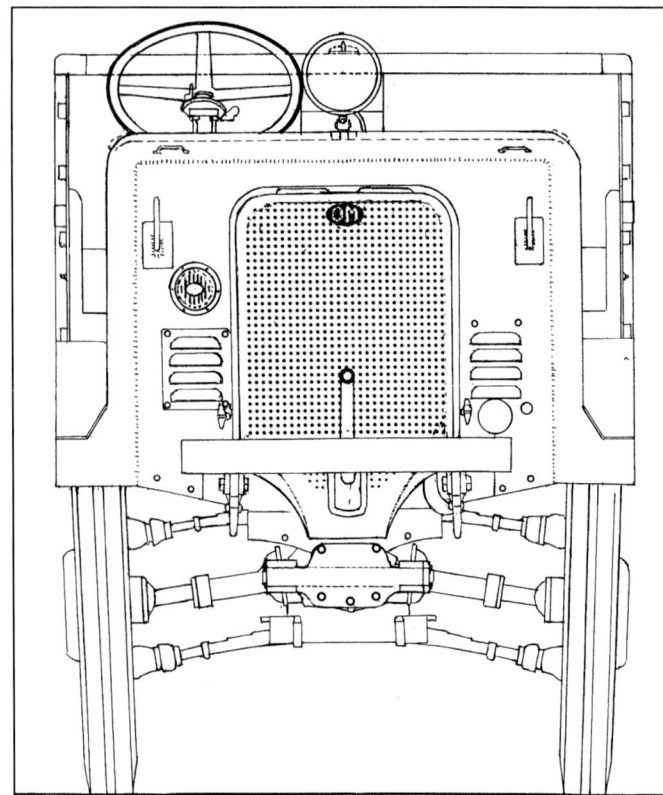
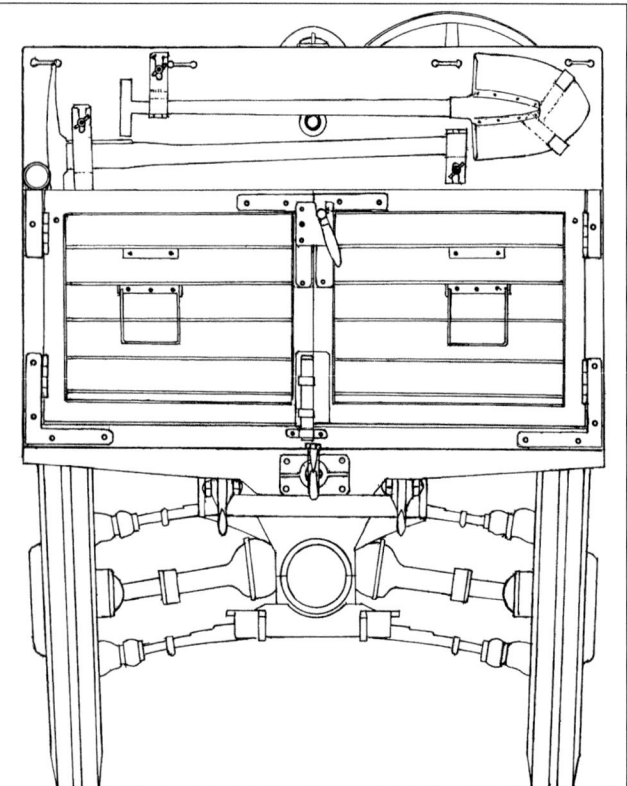

The Autocarretta 35. (Drawings by N. Pignato via GMT)

The *autocarrette* had their baptism of fire in Spain within the *Italian Corpo Truppe Volontarie*, or Voluntary Troops Corps. The camouflage usually included the canvas top, not deployed in the open position here. The local number plate, not always applied, was probably with blue background with yellow letters and numbers. The code 'BS' stands for Base Siviglia (Base Seville). (AUSSME)

road holding, due to the adoption of more efficient steering and transmission components. The model 35, together with the 32 retrofitted with the same modifications, would prove successful in operations in Italian East Africa and Spain in the late 1930s.

Based on field tests with vehicles assigned to a motorised division, a third series was developed, the Autocarretta 36, referred to as the 36 P (personale) – unofficially 36 DM (Divisione Motorizzata) – or 36 Mt (Materiale), depending on the configuration as a troop carrier (P) or a cargo carrier (Mt).

The final series was the Autocarretta 37, adopted shortly before the war, which had the same characteristics as the model 36 Mt but which reverted to semi-pneumatic tyres and lacked a windshield, as well as having minor modifications to the engine and other components.

The Autocarretta 35 and 36 were issued on varying scales to motorised divisions and to the Italian alpine divisions. They served in Italian East Africa (1,366 vehicles were operational as of 30 April 1936) and in Spain during the Civil War (many of which returned from the *Africa Orientale Italiana*); indeed, in 1937, of a total requirement for 9,640 of these vehicles, only 1,800 were available.

In October 1939, on the eve of Italy's entry into the Second World War, 2,751 autocarrette of various models were available on the national territory (excluding the colonies) with another 2,000 on order. They saw service in the Balkans, on the Eastern

The Autocarretta 36 P personnel carrier with Superflex Artiglio pneumatic tyres. Note the mounts for the machine guns and the absence of doors, replaced by chains. The circular bronze badge is also clearly visible.

The Autocarretta 36 P could carry a rifle squad, of commander, six fusiliers and two machine gunners with two Breda 30 light machine guns, plus the driver.

Front and in North Africa. Production of the Autocarretta series ceased in 1942, by which time the new SPA CL 39 was being issued; more than 5,000 of all series were produced.

The Autocarrette 35 and 36 enjoyed some degree of export sales: in 1940 a number of the 36 P version were sold to Portugal, following sales of unspecified numbers of the model 35 to Spain and 36 P to Ecuador; in 1942 100 of the model 36 were sold to Hungary.

Technical Description

The OM series of autocarrette were produced in a number of different series and configurations, mainly marked by changes in body styles. All these were characterised by an open cab mounted over the engine, although a retractable canvas top rolled up behind the driving compartment could be used to cover the top of the compartment.

The Autocarretta was the first four-wheel drive vehicle (except the artillery tractors) to be adopted by the *Regio Esercito*; all four wheels steered. The Autocarretta 35, like the 32 from which it derived, had a wooden body and was intended as a cargo carrier; the model 36 Mt likewise had a wooden cargo body, whereas the 36 P (or 36 DM) had a metal body consisting of three transverse rows of benches that enabled ten passengers (the driver plus an armed rifle squad, of a commander, two machine gunners and six fusiliers) and two Breda 30 light machine guns, or one Breda 37 heavy machine gun, to be carried. Both versions of the model 36 had a retractable canvas top that could be erected over the troop and driver's compartments.

The wheels on the Autocarretta 35 were stamped steel disc wheels with ten lightening holes (five large and five small, alternating) with semi-pneumatic tyres; the wheels on the 36 were more modern-looking stamped steel disc wheels with six lightening holes (four small and two somewhat larger) fitted with pneumatic tyres. The semi-pneumatic tyres could receive

The Autocarretta 36 Mt cargo carrier, without and with protective canvas cover.

traction improving devices made of metal tyre shoe attachments. All four wheels had independent suspension, with semi-elliptical leaf springs. The brakes were drum brakes on all four wheels, with the hand brake acting on the transmission. The electrical system was based on a dynamo, but starting was by means of a hand crank. The engine was a four-cylinder gasoline engine developing 23 HP; it was cooled by forced air (a feature studied for use in very cold climates, like the Alps), each separate cylinder having cooling fins. The transmission had four forward and one reverse speeds and a reduction gear. The autocarrette

The Autocarretta 36 Mt, with Superflex Stella Bianca pneumatic tyres. The rear hatch is shown in the two positions closed and open.

The Autocarretta 37 during factory trials. (OM)

This Autocarretta 37 under test is fitted with semi-pneumatic tyres and transports various wheels with pneumatic tyres and Artiglio treads. External tools are absent. (OM)

were equipped with a backstop device, acting on the primary shaft of the gearbox, useful in uphill starts to avoid the use of the brakes. There was also a mechanical stabilisation system, designed to dampen the vertical oscillations of the front differential.

Variants

On the cargo bed of some Autocarretta 35s a sprinkling system for ground decontamination was installed – the Attrezzatura Irroratrice modello 33. Additionally at the same time a smoke-producing system was manufactured – the Attrezzatura Nebbiogena Modello 33. In summary, this latter consisted of a specially made drum of about 300 litres capacity containing the smoke liquid to be sprayed, a cylinder of compressed-air to provide pressure and a series of pipes with atomising nozzles. Both these special vehicles were assigned to chemical units. Later, for the Autocarretta 32 and Autocarretta 36 a very similar smoke-producing system was made – the Attrezzatura Nebbiogena Modello 39.

On a small but unspecified number, perhaps 20, of chassis of the model 36 an armoured superstructure was mounted, becoming the Autocarretta Ferroviaria Blindata Modello 1942 (armoured railway truck). Some vehicles of this kind were used

The smoke-producing system Attrezzatura Nebbiogena modello 39 could be installed on the Autocarretta 32 and 36. (ACS)

in Dalmatia and Slovenia from December 1942 for the control and safety of narrow gauge (76cm) railway lines, and some were used by the Germans after the Armistice.

Specifications
- Designation: Autocarretta OM 32, 35 and 36
- Producer: Officine Meccaniche (OM), Brescia
- Years produced: 1932–1942
- Number produced: more than 5,000
- Length: 3,770mm (12ft 4in) model 35; 3,910mm (12ft 10in) model 36 Mt; 4,170mm (13ft 8in) model 36 P
- Width: 1,300mm (4ft 3in) model 35; 1,420mm (4ft 8in) model 36 Mt and 36 P
- Height: 2,200mm (7ft 3in) model 35; 2,100mm (6ft 10in) model 36 Mt and 36 P
- Unladen weight: 1,580kg (3,483lbs) model 35; 1,600kg (3,527) model 36 Mt; 1,650kg (3,638) model 36 P
- Carrying capacity: 800kg (1,764lbs) model 35 and model 36 Mt; 10 armed soldiers model 36 P
- Wheelbase: 2,000mm (6ft 7in)
- Front track: 1,100mm (3ft 7in) model 35; 1,070mm (3ft 6in) model 36 Mt and 36 P
- Rear track: 1,100mm (3ft 7in) model 35; 1,070mm (3ft 6in) model 36 Mt and 36 P
- Minimum turning radius: 3,500mm (11ft 6in) model 35; 4,200mm (13ft 9in) model 36 Mt and 36 P
- Minimum clearance: 450mm (1ft 6in)
- Bed, internal length: 1,700mm (5ft 7in)
- Bed, internal width: 1,200mm (3ft 11in)
- Bed, internal height: 500mm (1ft 7in)

Autocarretta Ferroviaria Blindata modello 1942 (armoured railway truck) made on the model 36 chassis.

- Towing capacity: 1,000kg (2,204lbs)
- Tyres: Celerflex 120 x 670 model 35 and model 37; Superflex Artiglio or Stella Bianca 7.00 x 18 model 36 Mt and 36 P
- Engine: AM, four-cylinder in-line, forced air-cooled, 1,616cc, 20 HP @2,400 rpm
- Fuel: Gasoline
- Transmission: four speeds forward, one reverse, with reduction gear
- Fuel capacity: 35 litres (9.2 US gallons, 7.7 Imperial gallons)
- Drive layout: 4x4
- Maximum speed: 25km/h (15.5mph) model 35; 45km/h (28mph) model 36 Mt and 36 P; 36km/h (22.3mph) model 37
- Range (on road): 160km (99 miles)

Autocarro leggero SPA 25 C

The SPA 25C still had the structure of a First World War truck.

Developmental and Service History

The SPA 25 C was developed as a civilian truck, in two versions with different engines: the 25 C/10 and the more powerful 25 C/12.

With the end of production of the Fiat 15 ter light military truck in 1922, the Italian military authorities urgently needed a replacement light truck in the ambulance configuration. The SPA 25 C/10 version was found to be suitable as a replacement, and in 1924 the *Regio Esercito* acquired a first batch of chassis with ambulance bodies fitted by Garavini of Turin. Increasing numbers of the 25 C/10 were then acquired in a number of different body styles. The last series produced had a closed cab with full doors, but without windows, a different radiator and a redesigned bonnet.

A SPA 25 of the last series produced (the photo is from 1940). The bodywork now features a closed cab with full doors but no windows. (ACS)

SPA 25 C/10 ambulance used during the Spanish Civil War (1938). It mounts new disc wheels with pneumatic tyres.

Close-up view of a SPA 25 C/10 ambulance. Note the logistics lettering stencilled in white on the side of the cab.

The 25 C/10 served in various mainly secondary roles throughout the Italian campaigns in East Africa and Spain, and the Second World War. Many examples of SPA 25C were used by the *Regia Aeronautica*. Poland also took delivery of 375 SPA 25 C trucks based on a 1924 order.

Technical Description

The SPA 25 C/10 was a 4x2 truck. The cab of the truck version was an open cab that had a removable canvas top; the ambulance version had a cab with a closed hard top with half-doors and canvas side curtains. Successive versions saw modifications to fender, radiator and bonnet styles. The two-axle truck was a rear-wheel drive vehicle. Semi-pneumatic tyres were mounted on pressed steel disc wheels, dual on the rear axle; later they were replaced by pneumatic tyres. Drive was right-hand. The frame was a ladder type and suspension consisted of semi-elliptic leaf springs; brakes were mechanical drum brakes. Some vehicles had an electrical system with a battery for lights and self-starting, while most apparently used a hand crank for starting. Following common practice for Italian military trucks, acetylene emergency lamps were fitted on the 25 C, in addition to the electric headlights.

The engine was a four-cylinder gasoline engine developing 39 HP and was fitted with a Zenith carburettor. The manual transmission consisted of four forward speeds and one reverse.

The truck bed was wooden with fixed sides and had a hinged tailgate. A canvas top was provided for the bed. There were small storage lockers on the running boards on both sides, and the spare tyre was mounted on a post which in turn was mounted on the frame next to the driver's compartment.

Variants

In addition to the basic cargo body, variants included passenger bus, ambulance, mobile workshop, refrigerated truck and tank truck versions. The truck was also built in a version with a larger engine, designated the SPA 25 C/12, small numbers of which were used by the military.

On the SPA 25 C/10 a smoke generator was mounted consisting of two tanks of 350 litres of liquid, a cylinder of compressed-air and a diffusion system. Although ready in 1935, this vehicle was rarely used, supplanted in the role by the OM 35 and 36 trucks.

Specifications

- Designation: Autocarro Leggero SPA 25 C
- Producer: Società Piemontese Automobili (SPA), Turin
- Years produced: 1925–1934
- Number produced: Not Known
- Length: 5,700mm (18ft 8in)
- Width: 1,800mm (5ft 11in)
- Height: 2,900mm (9ft 6in)
- Unladen weight: 2,300kg (5,071lbs)
- Carrying capacity: 2,500kg (5,512lbs); ambulance version 1,800kg (3,968lbs)
- Wheelbase: 3,250mm (10ft 8in)
- Front track: 1,510mm (4ft)
- Rear track: 1,510mm (4ft)
- Minimum turning radius: 6,750mm (22ft 2in)
- Minimum clearance: 350mm (1ft 2in)
- Bed, internal length: 2,900mm (9ft 6in)
- Bed, internal width: 1,680mm (5ft 6in)
- Bed, internal height: 700mm (2ft 3in)
- Tyres: semi-pneumatic tyres 895 x 135; pneumatic tyres Cord 32 x 6
- Engine: SPA C10 four-cylinder water-cooled, 2,700cc, 39 HP @2,500 rpm
- Fuel: Gasoline
- Transmission: four speeds forward, one reverse
- Fuel capacity: 80 litres (21 US gallons, 17.6 Imperial gallons)
- Drive layout: 4x2
- Speed: 50km/h (31mph)
- Range (on road): 290km (180 miles)

Autocarro Leggero Militare SPA 38 R and 36 R

Developmental and Service History

In early 1933 the *Regio Esercito* asked Fiat to develop a new 2-ton truck to replace the SPA 25 C. Two versions were ordered, one with a liquid-cooled engine and the other with an air-cooled engine. The two prototypes were ready in early 1934 and were subjected to a series of trials. Having met the required specifications, both truck models were adopted in early 1935; production was undertaken by SPA, a Fiat subsidiary.

A factory-new SPA 38 R, complete with canvas top. (Fiat)

Left side, without canvas top. (MSGR)

SPA 38 R trucks ready for delivery. (Fiat)

The SPA 36 R. The differences are evident in the bonnet and radiator grille. (Fiat)

SPA 38 R ambulance and bus civilian configurations. (MSGR)

The truck with liquid-cooled engine was named 38 R and was more widely issued. The air-cooled 36 R, was instead envisioned for use in desert and mountain environments; it was produced in limited numbers because field use in Italian East Africa in 1937 and in Libya in 1938 revealed that it was subject to frequent breakdowns, highlighting poor overall reliability. The 38 R, on the other hand, had participated with the Italian contingent in the Spanish Civil War and proved to be a solid truck with good performance on roads, although somewhat less so cross-country. In Spain it was used to tow infantry 65/17 guns, as well as acting as a prime mover for field and heavy field artillery. It saw extensive use in the Second World War in a variety of roles.

SPA 38 R military ambulance.

A SPA 38 R belonging to the Italian *Corpo Truppe Volontarie* in Spain. The camouflage, that extends to the canvas cover and licence plate type, is also visible on other vehicles featured in this study. (AUSSME).

The 38 R, whose configuration was similar to that of most Italian military trucks of the early 1930s, was a simple, robust machine that was easy to drive and performed well over rough ground. These qualities induced the French to order 500 examples for the *Armée de l'Air* (the French Air Force). About 400 vehicles were delivered before the war; the French used them as prime movers for the 75mm gun. Throughout the war, the truck was used in large numbers by the *Regio Esercito*, both in the standard version as well as the so-called colonial version, which had an oil-bath air filter, battery and a new electric starter motor, a 100-litre fuel tank, power take-off and other features. The *Regia Aeronautica* adopted a slightly dissimilar version of the truck, the model 38 RA, with a wider wheelbase (3,600mm), different rear brakes and other minor changes; an ambulance version based on the 38 RA was also used by the *Regia Aeronautica*. During the post-war period, production resumed of the 38 R/45, which had a battery, an 88-litre fuel tank, and a new completely enclosed cab. The 38 R remained in service in the Italian Army for many years after the end of the war.

Technical Description
The 38 R had a conventional 4x2 layout with a standard cab configuration. It was a two-axle truck with rear-wheel drive and dual rear wheels. The cab doors were half-doors; canvas side panels were provided to close off the cab in inclement weather. As with most Italian military trucks of the period, it had right-hand drive. The ladder frame consisted of two steel side frames and seven cross-members. The suspension consisted of leaf springs.

A SPA 38 R with a Red Cross flag stationary in the desert of Sidi Barrani in September 1940. (ACS)

An example bogged down in the Ukrainian mud; the soldiers attempting to push the truck have spread branches on the ground in an attempt to provide some traction. (AUSSME)

The engine was the Fiat 18 R which was also used on the Fiat Dovunque truck. The 18 R was a four-stroke gasoline engine assisted by a Weber 42 AK carburettor. The engine was mounted on guide rails that allowed it to slide forward for repairs and inspection. Ignition was by a Marelli FL4 magneto. The 38 R Coloniale had instead a Marelli A 20/12 electric starter and two packs of Marelli 3MF15 6-volt batteries. Cooling was by a centrifugal pump and a fan; the radiator consisted of eight elements

A SPA 38 R carrying a contingent of Bersaglieri in North Africa. (AUSSME)

Two 38 Rs equipped with Breda 20mm machine guns escort a column led by a Fiat 626. Eastern Front, summer 1941. (ACS)

which could be closed off from each other. The manual transmission, which was not synchronised, had four forward and one reverse gears. Drum brakes were present on all wheels. The truck bed was wood with fixed sides and had a hinged tailgate with mounting steps. Five wooden benches could be fitted to the bed to carry troops. A canvas top with five bows was fitted. The spare tyres were stored below the body, under storage lockers.

This accident involving a SPA 38 allows the chassis to be seen from below.

The mobile repair shop Modello 37 on the SPA 38 R chassis. Viberti designed a similar model.

Variants

In addition to the basic truck, variants included a refrigerated truck, ambulance, mobile repair shop, field office, passenger bus, and at least one mobile library.

The 20mm Breda anti-aircraft cannon was often mounted on the bed of the 38 R.

A SPA 36 R, with its characteristic oval radiator grille, precedes a 38 R.

Specifications

- Designation: Autocarro Leggero Militare SPA 38 R/SPA 36 R
- Producer: Società Piemontese Automobili (SPA), Turin
- Years produced: 1935–post-war
- Number produced: Not Known
- Length: 5,783mm (18ft 11in) 38 R; 5,830mm 36 R (19ft 2in)
- Width: 2,070mm (6ft 10in) 38 R; 2,000mm (6ft 7in) 36 R
- Height (with canvas cover): 2,780mm (9ft 2in)
- Unladen weight: 3,200kg 38 R; 3,250kg 36 R
- Carrying capacity: 2,500kg (5,512lbs)
- Wheelbase: 3,500mm (11ft 6in)
- Front track: 1,545mm (5ft)
- Rear track: 1,427mm (4ft 8in)
- Minimum turning radius: 5,700mm (18ft 8in)
- Minimum clearance: 250mm (10ft)
- Fording depth: 600mm (2ft)
- Bed, internal length: 3,200mm (10ft 6in)
- Bed, internal width: 1,980mm (6ft 6in)
- Bed, internal height: 670mm (2ft 2in)
- Tyres: Cord 32 x 6; Ultraflex 210 x 20
- Engine: Fiat 18 R, 4,053cc, water-cooled, 56 HP @2,000 rpm 38 R; Fiat six-cylinder air-cooled, 4,426cc, 50 HP @2,000 rpm 36 R
- Fuel: Gasoline
- Transmission: four speeds forward, one reverse
- Fuel capacity: 38 R, 108 litres (28.5 US gallons, 23.8 Imperial gallons); 36 R, 100 litres (26.4 US gallons, 22 Imperial gallons)

LIGHT TRUCKS 115

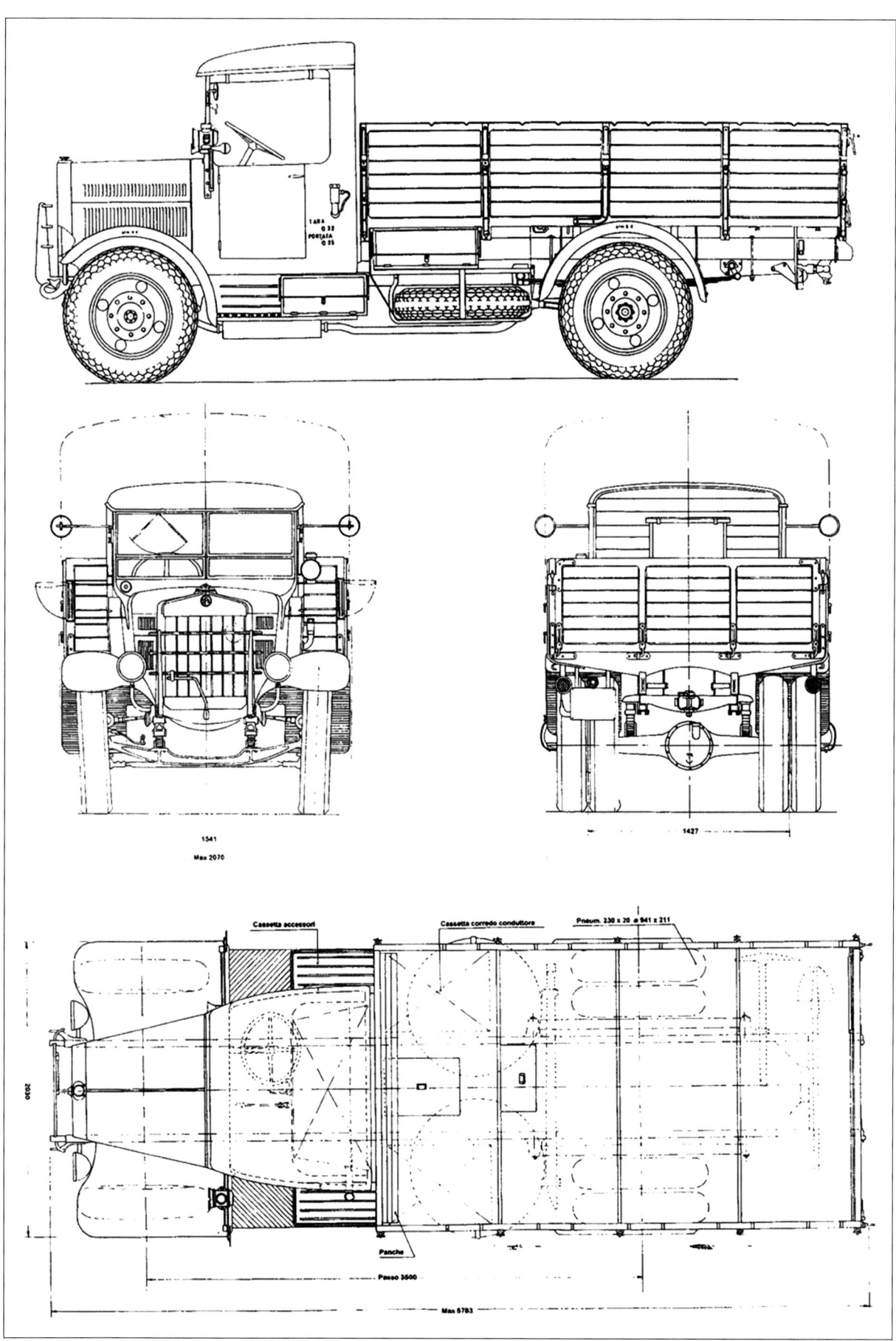

Line drawing of the SPA 38 R

- Drive layout: 4x2
- Speed: 52km/h (32mph)
- Range (on road): 310km (192.5 miles)
- Range (cross-country): 290km (180 miles)

Autocarro Leggero Fiat 618 MC

The Fiat 618 Militare Coloniale light truck. (Fiat)

A Fiat 618 MC in East Africa.

Men of the 6ª *Divisione Camicie Nere 'Tevere'* of the MVSN, in Somalia in the second half of the thirties.

A Fiat 618 of the Italian CTV corps in Spain.

Fiat 618, possibly destroyed during the battle of Santander in Spain, in August 1937. It belonged to *724th Bandera Inflessibile* of the *7th Gruppo Banderas*, part of the *2nd Fiamme Nere* Volunteer Division of the MVSN. Each *Bandera* of the MVSN corresponded to a battalion and the *Gruppo Banderas* to a regiment.

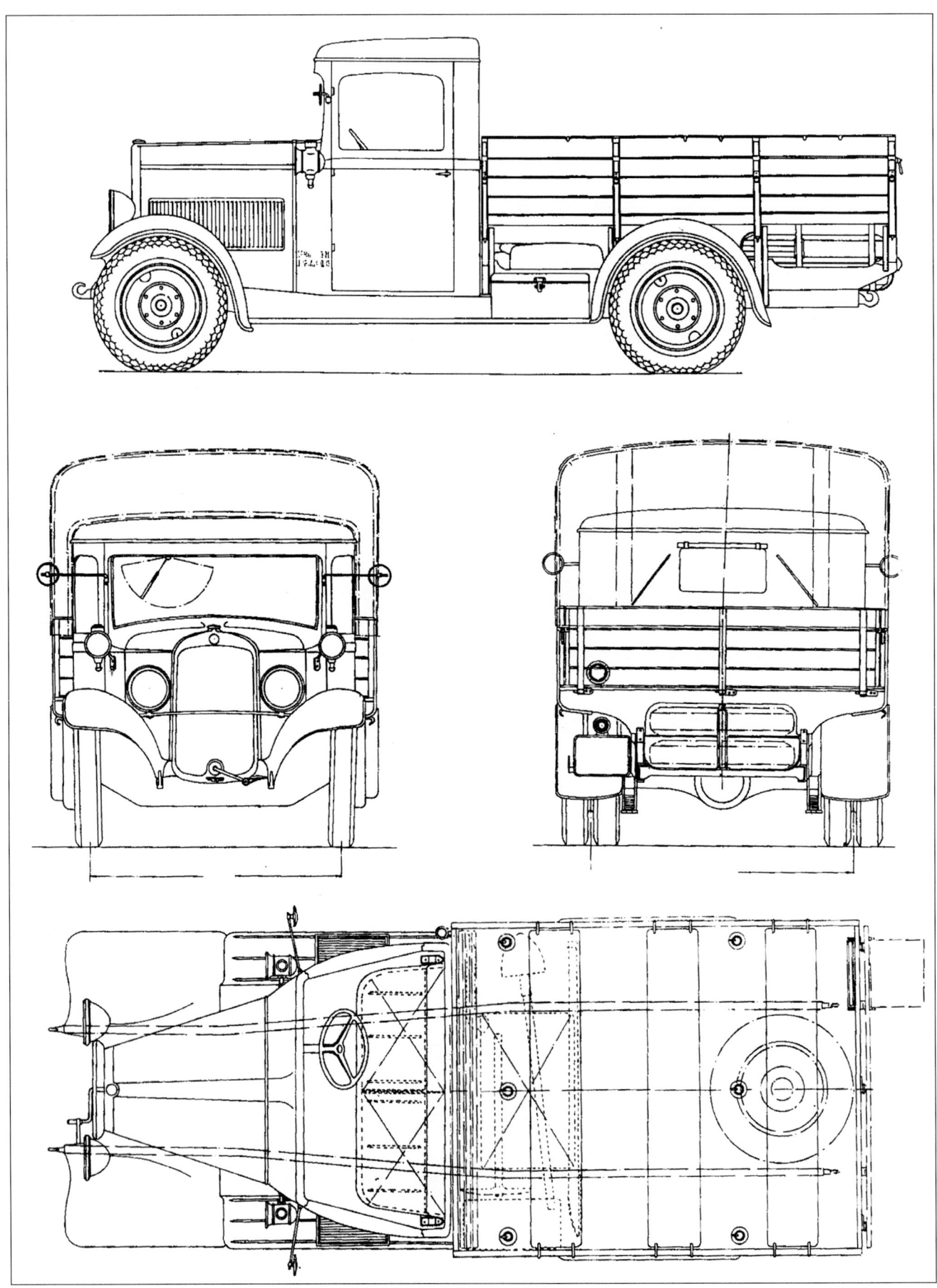

Line drawing of the Fiat 618, August 1935. (Fiat)

The ambulance bodywork made by Viberti.

Developmental and Service History

The Fiat 618 Militare Coloniale, more commonly referred to as simply the 618 MC, was the military version of the civilian Fiat 618, designed for light transportation duties that had entered production in 1934. More precisely, it was derived from modified version 618 C (Coloniale), which entered service in 1935 and was conceived for use in Italian North and East Africa. In these territories, it proved to have several problems, including engine overheating, a weak transmission, problematic suspension and limited cargo capacity when operating in the mountains. Eventually, despite its widespread use, the truck was relegated to duties as a squad carrier for infantry units.

Nevertheless, the 618 MC saw extensive service in Ethiopia and Somalia in 1935–1936 (probably more than 3,000 examples), and later almost 1,700–2,000 were assigned to the Italian *Corpo Truppe Volontarie* (the CTV, the Italian Volunteer Corps) that served in the Spanish Civil War. In mid–1937 each 37mm anti-tank battery, equipped with the German 3.7cm Pak 36 gun, in the CTV in Spain was issued with six Fiat 618 MC trucks as prime movers for the guns. Later that year, a number of 618 MC trucks were fitted with St Etienne machine guns for anti-aircraft defence duties. The next year, following exercises carried out in Libya, the truck was once again judged not sturdy and was thus considered suitable only for light duties.

The Fiat 618 MC was used predominantly by the *Regio Esercito*, but other users included the *Milizia Volontaria per la Sicurezza Nazionale* (MVSN), the Finance Guard, the Public Security Corps and the *Regia Aeronautica*. The Fiat 618 was manufactured under licence in Poland for both the civilian and the military market. Although its production in Italy was terminated in 1937, being replaced by the CL 39, the 618 MC continued to serve throughout the war in lieu of more suitable vehicles.

Technical Description

The Fiat 618 MC was a conventionally laid out 4x2 truck with rear-wheel drive and dual rear wheels. The closed cab with full doors had a wooden frame with iron reinforcements, covered with sheet metal. The windshield was in one piece and could be opened forward. A manual windshield wiper on the right side, the driver's side, and rear-view mirrors were optional accessories. The cargo body was made of wood, with a one-piece tailgate. Soldiers on board could sit on three removable benches. Pneumatic tyres were mounted on pressed steel wheels that had two lightening holes per wheel; two spare wheels and tyres were carried under the rear of the cargo bed. Brakes were hydraulic drums on all four wheels. The hand parking brake acted on the transmission. The electric lighting system used a dynamo and no battery was present; the engine was started by means of a hand crank. The engine was a four-cylinder gasoline engine which developed 43 HP; the transmission had four forward and one reverse gear.

Variants

Variants of the 618 MC included a radio van, ambulance and passenger bus.

Specifications
- Designation: Autocarro Leggero Fiat 618 Militare Coloniale
- Producer: Fabbrica Italiana Automobili Turin (Fiat), Turin
- Years produced: 1934–1937
- Number produced: Not Known
- Length: 4,700mm (15ft 5in)

- Width: 1,940mm (6ft 4in)
- Height: 2,500mm (8ft 3in)
- Unladen weight: 2,115kg (4,662lbs)
- Carrying capacity: 1,200kg (2,645lbs)
- Wheelbase: 3,050mm (10ft)
- Front track: 1,490mm (4ft 11in)
- Rear track: 1,540mm (5ft)
- Minimum turning radius: 5,750mm (18ft 11in)
- Minimum clearance: 200mm (8in)
- Tyres: Superflex 6.00 x 20
- Engine: Fiat model 118A, four-cylinder in-line water-cooled, 1,944cc, 43 HP @3,800 rpm
- Fuel: Gasoline
- Transmission: four speeds forward, one reverse
- Fuel capacity: 60 litres (15.8 US gallons, 13.2 Imperial gallons)
- Drive layout: 4x2
- Maximum speed: 65km/h (40mph)

FIAT 508 M CAMIONCINO AND FURGONCINO

The Fiat 508 Camioncino three-speed, left-hand drive.

The Fiat 508 Furgoncino three-speed, right-hand drive.

Developmental and Service History

In 1932 the *Regio Esercito* adopted a compact car, the Fiat 508 M, which proved to be well-liked and successful (see above). Subsequently, a small pick-up truck referred to in Italian as a 508 Camioncino was developed based on the 508 A and 508 B chassis (three-speed and four-speed respectively), as was a small, enclosed van, the 508 Furgoncino, for both the civil and military markets.

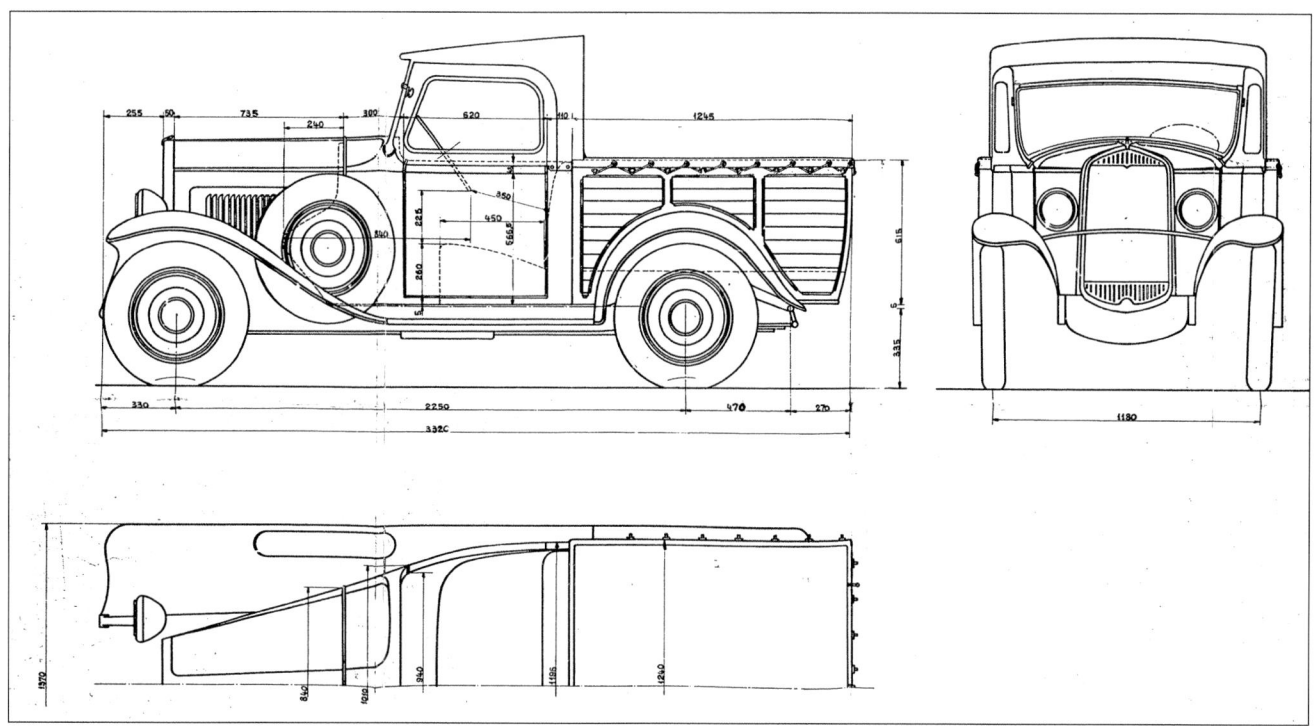

Line drawing of the 508 Camioncino three-speed, left-hand drive, December 1931. (Fiat)

The Fiat 508 Camioncino four-speed, recognisable by the different shape of the grille and the engine hood.

Two Camioncini four-speed trucks in Asmara, Eritrea, in the second half of the 1930s.

A 508 Camioncino four-speed, without spare wheel, in Asmara in 1943. (V. Cerenzia / www.acrinews.it)

The civilian Fiat 508 Camioncino four-speed, with a slightly different body.

These small trucks served the *Regio Esercito* throughout the war, and were also used by the *Regia Marina* and the *Regia Aeronautica*, seeing heavy use (at least 497 Camioncini and 20 Furgoncini) in Italian East Africa.

Technical Description

The 508 Camioncino and Furgoncino shared almost all of their mechanical components with the 508 A/B/M passenger car on which they were based. The Camioncino had a metal cab with half-doors, a canvas top and rear panel. The Furgoncino had an all-metal cab and body and had full doors. The windshield was of a one-piece type. The bed, bed sides and tailgate of the Camioncino were made of wood, with folding benches for transporting troops, while the body of the Furgoncino, as noted, was all metal. Pneumatic low pressure tyres were mounted on pressed steel wheels and there were two spare wheels/tyres mounted on the sides of the bonnet. All four wheels had drum brakes. Suspension was leaf springs front and rear with enhanced hydraulic

The Fiat 508 Furgoncino four-speed. (Fiat, ACS)

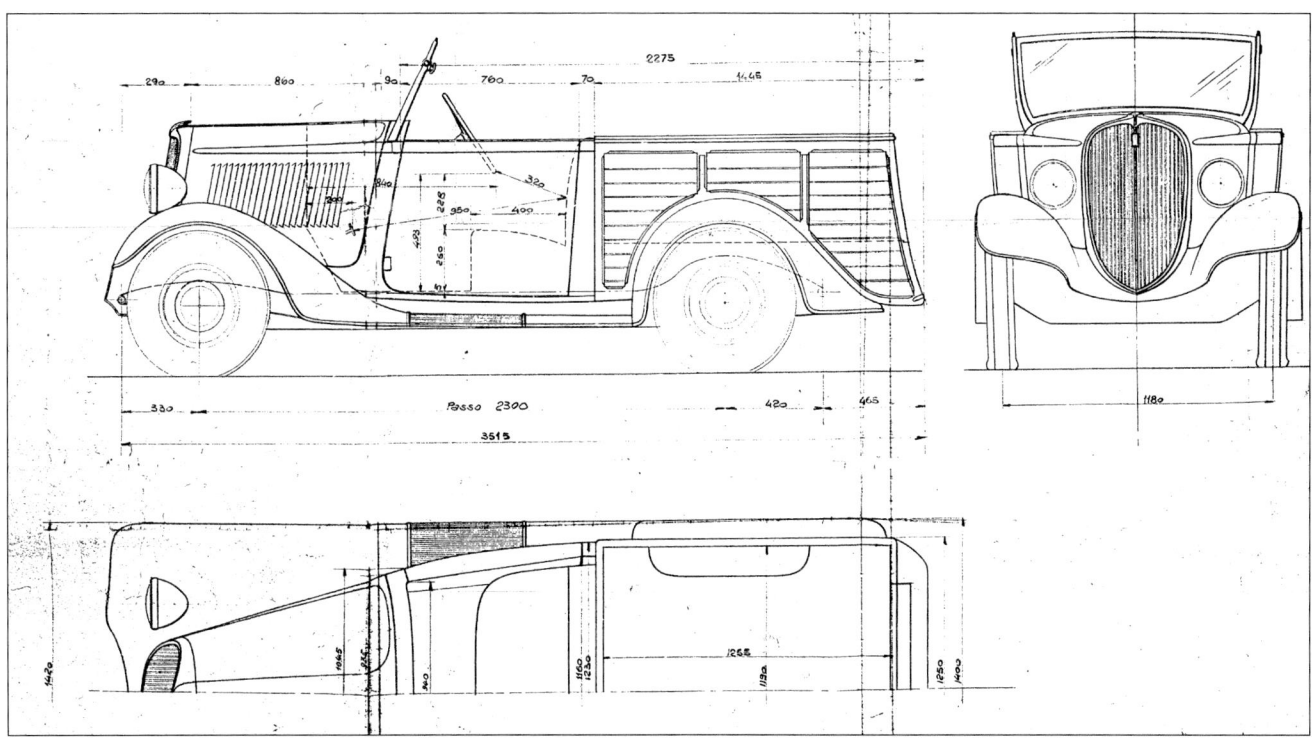

Bodywork design for a 'Camioncino 508 III' dated 20 September 1934. The vehicle is based on the four-speed 508 B, but with some elements (mudguards and cargo bed with a 'protruding tail') that anticipate the future 508 L Camioncino. (Fiat)

shock absorbers compared to civilian models. The engine was a four-cylinder gasoline engine developing 20 HP. The manual transmission of the first version had three forward and one reverse gears, while the second version and variants had four forward and one reverse gears. The military version could have an oil air filter instead of a dry one.

Variants

On a few dozen examples of Camioncino (50 or 68, depending on the sources), belonging to 508 M and 508 C / 1100 version, a weapon system of two coupled 8mm Fiat 35 machine guns was mounted.

On the long wheelbase Furgoncino a small ambulance could be set up, the body style differing depending on the coachbuilders.

Specifications

- Designation: Fiat 508 M Camioncino; 508 M Furgoncino
- Producer: Fabbrica Italiana Automobili Turin (Fiat), Turin
- Years produced: 1933–1937
- Number produced: NA
- Length: 3,320mm (10ft 11in) first series; 3,515mm (11ft 6in) second series
- Width: 1,380mm (4ft 8in)
- Height: 1,450mm (4ft 9in) with top down; 1,600mm (5ft 3in) with top up (and pick-up version)
- Unladen weight: 715kg (1,576lbs)
- Carrying capacity: 350kg (772lbs) Camioncino; 300kg (661lbs) Furgoncino
- Wheelbase: 2,250mm (7ft 4in) first series; 2,300mm (7ft 6in) second series
- Front track: 1,800mm (5ft 9in)
- Rear track: 1,800mm (5ft 9in)
- Minimum turning radius: 4,750mm (15ft 7in)
- Minimum clearance: 220mm (9in)
- Tyres: Superflex 4.25 x 17
- Engine: Fiat 108 M four-cylinder, side-valve, water-cooled, 995cc, 20 HP @3,400 rpm first series (three-speed); 24 HP @3,800 rpm second series (four-speed)
- Fuel: Gasoline
- Transmission: three speeds forward, one reverse on first series; four speeds forward, one reverse (3rd and 4th gears synchronised) on second series
- Fuel capacity: 26 litres (6.9 US gallons, 5.7 Imperial gallons)

- Drive layout: 4x2
- Maximum speed: 65km/h (40mph)
- Range (on road): 310km (193 miles)

Fiat 508 C and 1100 Camioncino and Furgoncino

The Fiat 508 C Camioncino or 'Musetto' (small snout). (CSM)

The driver of a camouflaged 508 C Camioncino with a *Regio Esercito* number plate talking to a *Milizia della Strada* officer. (ACS)

Fiat 508 C Camioncino and Lancia 3Ro trucks in North Africa.

Budapest, 1941. A convoy with several Bianchi Miles trucks ready to head to the Eastern Front. In the foreground, a Fiat 508 C Camioncino with the symbol of the *Stato Maggiore Regio Esercito* (Army Staff). (ACS)

508 C Camioncini from the *Stato Maggiore Esercito* on the Eastern Front in 1941. (ACS)

This image is part of a series of poses taken on Agfa colour film by a German soldier during the transfer trip and stay in North Africa. On the left, an Opel Olympia 38 that appears camouflaged in the *Tropen* colours employed by the Wehrmacht since March 1942. On the right, a Fiat 508 C Camioncino belonging to the *Regio Esercito* whose grey-green colour is only partially covered by the *kaki sahariano* (sand-yellow) applied by spray gun.

The Fiat 1100 Camioncino or *Musone* (big snout). (Fiat)

A 1100 Camioncino used by the *Ufficio Propaganda Regio Esercito, Servizio Assistenza Truppe* (Army Propaganda Office, Troops Assistance Service).

Developmental and Service History

In the second half of the 1930s, the Italian armed forces incorporated a military version of the popular civilian car Fiat Nuova Balilla sedan, i.e. the third series of the Fiat 508, which was called the 508 C Militare (see above). As already done for the previous models, on its base a small pick-up truck version (*camioncino*) and an enclosed van (*furgoncino*) with an all-metal body were developed. The evolution of these two vehicles began from the exit from the market of the equivalents based on the old 508 M (see the Fiat 508 M Camioncino and the Fiat 508 M Furgoncino).

1100 Camioncino (*Musone*) crossing a pontoon bridge over the Donets, a river which traverses both Russia and the Ukraine. Its light colour, perhaps *kaki sahariano*, is interesting compared to the dark grey-green of the other vehicles. (ACS)

The 508 L Camioncino (small snout, long frame) was produced for the civilian market but also used by the *Regio Esercito*.

A 508 L Camioncino, painted grey-green.

The first series of the new pick-up truck was manufactured in two variants: the Fiat 508 C Camioncino (in this case, C stands for *corto*, i.e. short wheelbase) was produced until the end of the Second World War for military use only; the 508 L version (*lungo*, long wheelbase) was both civilian and military. The same was true for the 508 Furgoncino. Both of these small vehicles were adopted in 1939 and saw service throughout the war in all three branches of the Italian armed forces.

From 1941, both the Camioncino and the Furgoncino were produced on the basis of the restyled Nuova Balilla Fiat 508 C, better known as the Millecento or '1100'. As with the previous truck, the short wheelbase 1100 C variant was intended for military use only, while the 1100 L long wheelbase variant could be both civil or military.

Unusual rear view of a 508 L Camioncino.

A 1100 L Camioncino (big snout, long frame) captured by Commonwealth troops.

The *Istituto Luce* included a *Reparto Guerra* (War Detachment), with teams of cinematographers and photographers located with the various armed forces. Here a 1100 L Furgoncino in the spring of 1942. The vehicle has a civilian number plate. (ACS)

Technical Description

The 508 C / 1100 Camioncino was a small 4x2 pick-up truck. The body of the military variant was of mixed metal and wood construction: the bonnet, fenders, cowling, and cab roof were of metal, doors were a mix of metal and wood panels (long frame versions had all-metal doors and different mudguards), and the cargo body was wood. Compared to the sedan, the Camioncino frame was strengthened and the tyres mounted were larger. The earlier version based on the Nuova Balilla 508 C model was afterwards christened Musetto (small snout); this because the later version based on the new 1100 was nicknamed Musone (big snout), having a more prominent grille and a slightly different and less slanted bonnet.

In emergencies, these vans were used for other functions, as in this case for the transport of a wounded soldier on the Eastern Front in the summer of 1942. (ACS)

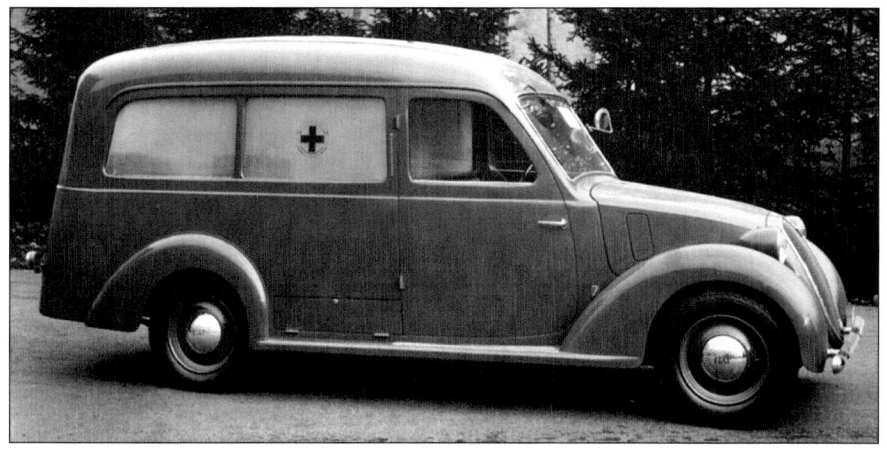

A civilian ambulance based on the 508 L frame. (Croce Verde Lugano)

The enclosed van version of the 508 C / 1100 Camioncino was known as the Furgone (van) or more commonly Furgoncino (small van); the body was all metal, similar to that on many small civilian delivery vans of the era. Because bodies were built by many independent coach builders there could be minor differences in body style and configuration. The independent front suspension on both types used coil springs and hydraulic shock absorbers; the rear suspension consisted of leaf springs, hydraulic shock absorbers and a stabiliser bar. Brakes were hydraulic drum brakes, plus a mechanical hand brake on the transmission. The engine was the same four-cylinder valve-in-head gasoline engine used on the 508 C sedan, developing 30 HP.

Variants

The Camioncino was built in a mobile repair van version. Some 508 M (the initial model), 508 C and 1100 Camioncino were also converted to armed versions, with armament consisting of twin Fiat 8mm model 35 air-cooled machine guns fed by box magazines, mounted on a modified 20mm gun mount.

Two types of ambulance body, one all-metal and the other of wood, were built on the lengthened frame of the Furgoncino of all models.

Specifications

- Designation: Fiat 508 Camioncino Militare; Fiat 508 Furgoncino Militare
- Producer: Fabbrica Italiana Automobili Turin (Fiat), Turin
- Years produced: 1939–1945 ca.
- Number produced: a few thousand Camioncini (508 M and 508 C / 1100 version) until 1943, 1,677 from September 1943 to 1945
- Length: 4,100mm (13ft 5in) 508 C Musetto; 4,115mm (13ft 6in) 1100 Musone

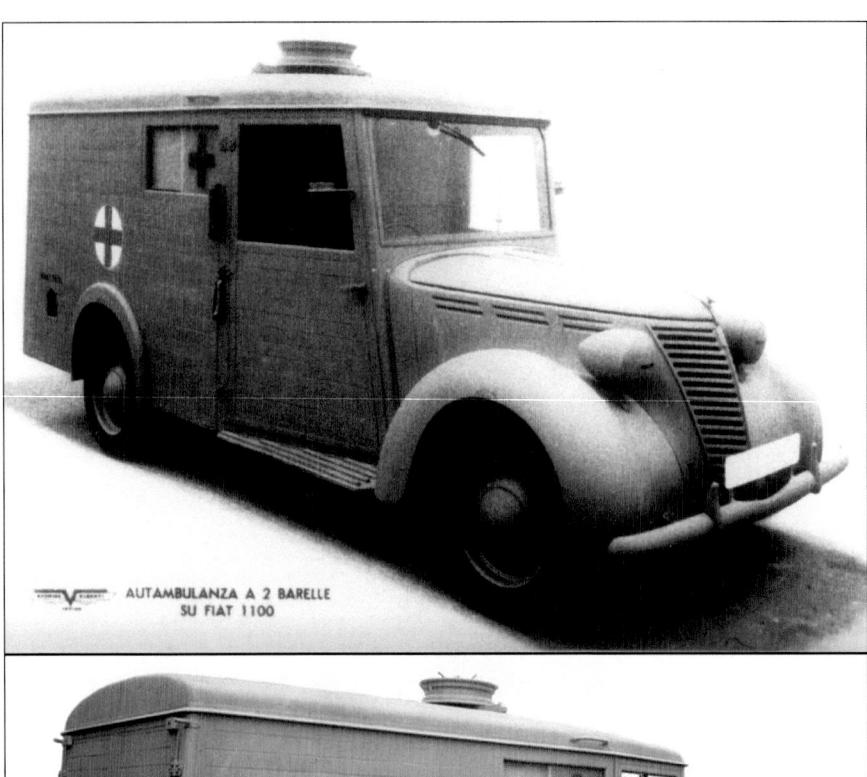

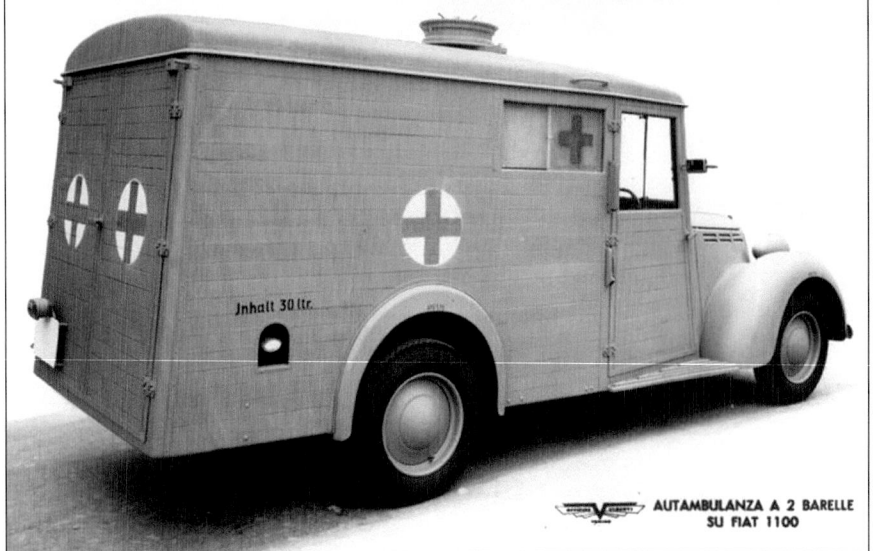

The Fiat 1100 two-stretcher ambulance with wooden bodywork. On the cab top a spare wheel without tyre is fixed. After the Armistice, the vehicle was produced for the Wehrmacht, as evidenced by the German wording next to the tank filler neck (*Inhalt 30 ltr.*, content 30 litres). (Viberti)

An example employed by the Germans, In this case the spare wheel is fitted with a tyre.

- Width: 1,520mm (5ft) 508 C Musetto; 1,660mm (5ft 6in) 1100 Musone
- Height without tarpaulin: 1,680mm (5ft 6in)
- Height with tarpaulin: 2,050mm (6ft 8in) 508 C Musetto; 1,530mm (5ft) 1100 Musone
- Unladen weight: 880kg (1,940lbs) 508 C Musetto; 1,008kg (2,222lbs) 1100 Musone
- Carrying capacity: 420kg (926lbs)
- Wheelbase: 2,427mm (7ft 11in) for *corto*; 2,707mm (8ft 10in) for *lungo*
- Front track: 1,265mm (4ft 2in) for *corto*; 1,316mm (4ft 4in) for *lungo*
- Rear track: 1,254mm (4ft 1in) for *corto*; 1,367mm (4ft 6in) for *lungo*
- Minimum turning radius: 5,000mm (16ft 5in)
- Minimum clearance: 230mm (9in)
- Bed, internal length: 1,540mm (5ft)
- Bed, internal width: 1,600mm (5ft 3in)
- Bed, internal height: 600mm (2ft)
- Tyres: Aerflex 6.00 x 16
- Engine: Fiat 108 C, four-cylinder water-cooled, 1,089cc, 30 HP @4,400 rpm
- Fuel: Gasoline
- Transmission: four speeds forward, one reverse
- Fuel capacity: 30 litres (8 US gallons, 6.6 Imperial gallons)
- Drive layout: 4x2
- Maximum speed: 75km/h (47mph)
- Range (on road): 270km (168 miles)

Autocarro Leggero SPA CL 39

Developmental and Service History

The CL 39 (initially designated the L39) was known or referred to by different names or designations: Autocarretta SPA (to distinguish it from the Autocarretta OM); the *Carro Leggero per Fanteria* (CLF, light infantry truck); Autocarro Leggero mod. 39, but the most commonly accepted designation is the Autocarro SPA CL 39.

The SPA CL 39 had its genesis in the requirement for a simple light infantry truck that could operate in Italy's mountain environment. It had to be less complicated and quicker to manufacture than the OM light trucks, as well as less expensive. The CLF, prototype of the CL 39, was ready in late 1938 and was adopted with some changes in 1939; specifications originally called for two versions of the CL, one with a water-cooled engine and one with an air-cooled engine, but the air-cooled option was discarded. A colonial version of the CL 39 followed in early 1941; it was characterised, in addition to the oil-bath filter, by larger tyres and the cargo bed being lower on the chassis. In addition to acting as an infantry squad carrier, the CL 39, which had good towing qualities, was used to tow a variety of weapons, from the 20mm Breda, to the ubiquitous 47/32 anti-tank gun, and even the larger 75/18 howitzer.

The CL 39 was produced in large numbers and in addition to its use primarily by the *Regio Esercito*, was also used by the *Regia Aeronautica*. In army use, it was assigned on a scale of six CL 39s to each 75/18 artillery group, eight machines to each 81mm

The prototype of the CLF 39 with obsolete, stamped 'artillery wheels' and Celerflex semi-pneumatic tyres. (F. Cappellano)

The SPA CL 39 with Celerflex semi-pneumatic tyres and Dayton-type spoked wheels.

The CL 39 with Superflex Artiglio tyres.

mortar company and four to each 47mm cannon company of the infantry divisions. Additionally, the artillery headquarters (regiment, group, battery) of the armoured divisions *Littorio* and *Ariete* each had three CL 39s. Other examples were assigned to the motorised divisions, both in standard and colonial versions. In 1943, 145 CL 39s were authorised for each infantry division.

Overall, the CL 39 was a very satisfactory and successful vehicle. It was easy to drive, easy to maintain, had excellent visibility, a tight turning radius, good suspension and good fuel consumption.

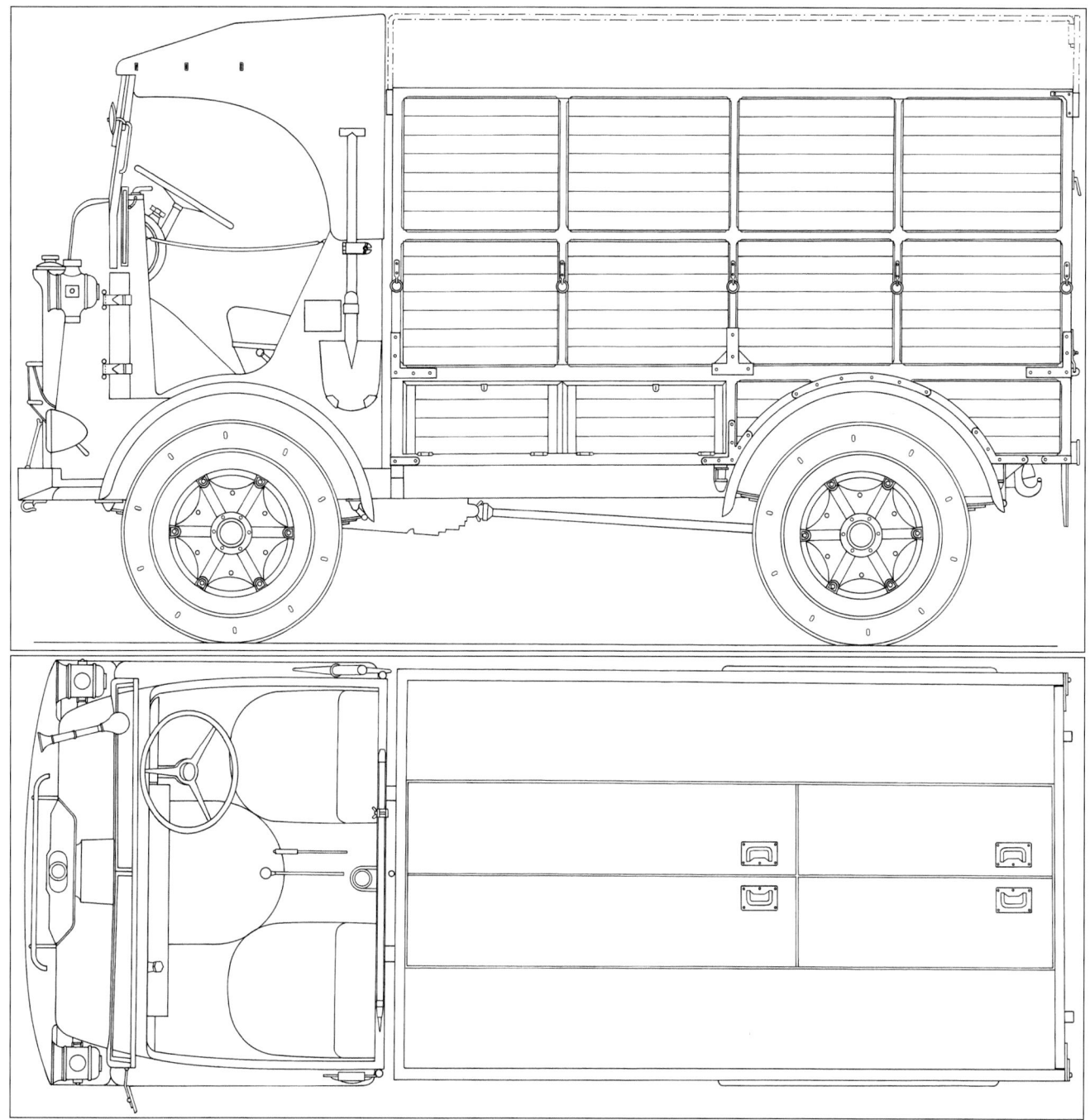

The CL 39 light truck with semi-pneumatic tyres. (Drawing by A. M. Feller – GMT)

The CL 39 saw service on all fronts in which the Italians were engaged, except for Italian East Africa. Following the 8 September Armistice, 198 CL 39s were delivered to German forces in 1944. Following the war, the CL 39 remained in service with the Italian Army until the 1950s.

Technical Description

The CL 39 had been designed as a very simple no-frills machine. It was a 4x2 truck with rear-wheel drive and a cab-over-engine configuration, fitted with the usual, for Italian trucks, right-hand drive. The ladder frame consisted of two rails with five transverse cross-members; a towing pintle was attached to the rearmost cross-member. The cab had a removable canvas top and canvas rear panel. Chain stitches were used in place of the doors. The wooden body had bench seats that could hold eight soldiers, folding back the benches allowed a 1,000kg payload to be carried. There were four storage lockers, two on each side, mounted below the bed. The bed of the metropolitan version of the CL 39 had high sides, whereas the sides on the colonial version were somewhat lower.

The wheels and tyres mounted on the CL 39 underwent an evolution: the earliest version had cast steel wheels with eight spokes and was fitted with Celerflex semi-pneumatic tyres; later versions had six-spoke wheels with 178mm wide Superflex

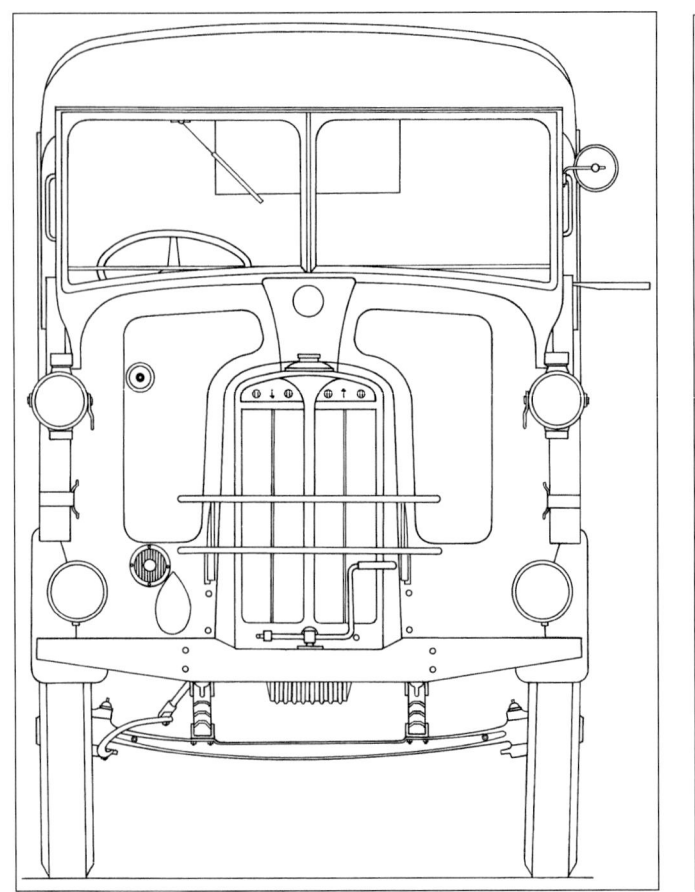
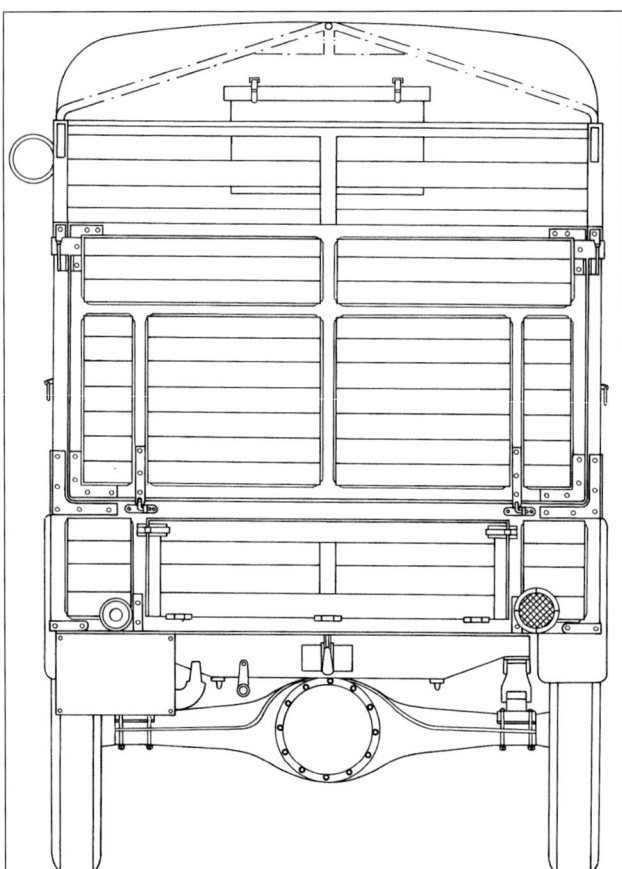

The CL 39 light truck with semi-pneumatic tyres. (Drawing by A. M. Feller – GMT)

The CL 39 Coloniale without and with canvas cover. The tyres are Ultraflex Sigillo Verde 210 x 18 balloon tyres.

Artiglio pneumatic tyres, except the colonial version which mounted the Ultraflex Sigillo Verde very low pressure tyres 210mm wide. The spare tyre on the metropolitan version was mounted beneath the rear of the bed, while the colonial version had the spare tyre mounted behind the cab. The oil-bath air filter of the colonial version was later extended to all versions. Suspension consisted of semi-elliptical leaf springs and shock absorbers. Brakes were hydraulic on all four wheels, while the hand brake acted on the transmission. The electrical system worked on a dynamo. The gasoline engine was a four-cylinder that developed 24 HP. The manual transmission, which was not synchronised, had five forward and one reverse gears; a reduction gear doubled the available speeds. The CL 39 also had a backstop device.

Variants

The CL 39 was built in a standard (metropolitan) version as well as a colonial version (CL 39 C), but changes throughout the production run often blurred the distinction between the two types.

A shower truck (Autobagno) was also manufactured, for decontaminating soldiers exposed to chemicals or gas, but finally used for personal hygiene.

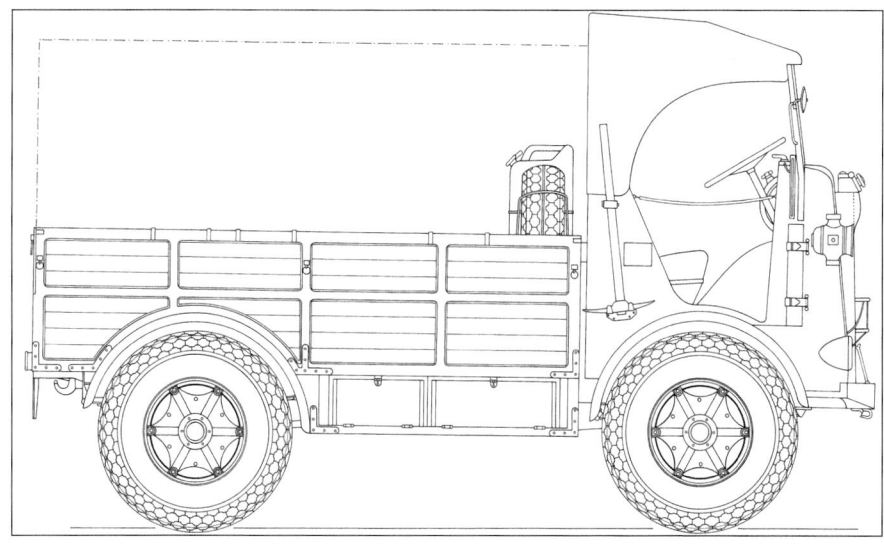

The CL 39 Coloniale. (Drawing by A. M. Feller – GMT)

A CL 39 from the *Regia Aeronautica* in North Africa in 1941. (Bortolotti)

A column of CL 39 light trucks in the Ukraine in 1941. The vehicles carry the regulation bronze vehicle badge and the number plate painted on the front bumper. (ACS)

Struggling with the ford of a watercourse, still in Ukraine. (ACS)

A CL 39 with semi-pneumatic tyres having just crossed a bridge built by Italian engineers over the Dnieper River; note the camouflage netting and the anti-aircraft weapon mounted in the bed. (ACS)

Italian soldiers in North Africa.

A CL 39 of the Italian *Regia Aeronautica* in Albania in winter 1940–1941. Note the different format of the number plate compared to that of the *Regio Esercito* seen earlier, and the absence of the circular bronze badge (ACS)

LIGHT TRUCKS 137

CL 39 light trucks of the *Giovani Fascisti* (the Young Fascists or GG.FF.) in Libya in the summer of 1941. The first vehicle is equipped with a Fiat 35 machine gun. Some clothing is used to protect the tyres from the heat of the sun. (Museo del Reggimento GG.FF.)

CL 39 shower truck.

Specifications
- Designation: Autocarro Leggero SPA CL 39
- Producer: Società Piemontese Automobili (SPA), Turin
- Years produced: 1939 to mid–1950s
- Number produced: 5,840
- Length: 3,890mm (12ft 9in)
- Width: 1,520mm (5ft)
- Height 2,300mm (7ft 6in)
- Unladen weight: 1,630kg (3,594lbs)
- Carrying capacity: 1,000kg (2,204lbs)
- Wheelbase: 2,300mm (7ft 6in)
- Front track: 1,300mm (4ft 3in)
- Rear track: 1,320mm (4ft 4in)
- Minimum turning radius: 5,000mm (16ft 5in)

Rare image of a CL 39 equipped with improvised armour used in Montenegro in September 1942; it probably belonged to the *XVIII Battaglione Mortai da 81mm* (a mortar battalion) of *18ª Divisione di Fanteria, 'Messina'*.

- Minimum clearance: 245mm (10in)
- Fording depth: 700mm (2ft 3in)
- Bed, external length: 2,480mm (8ft 2in)
- Bed, external width: 1,460mm (4ft 9in)
- Bed, external height: 995mm (3ft 3in)
- Tyres: Celerflex 140 x 620; Superflex Artiglio 7.00 x 18; Ultraflex Sigillo Verde 210 x 18
- Engine: SPA CLF, four-cylinder in-line, water-cooled, 1,628cc, 25 HP @2,400 rpm
- Fuel: Gasoline
- Transmission: five speeds forward, one reverse, with reduction gear
- Fuel capacity: 55 litres (14.5 US gallons, 12 Imperial gallons)
- Drive layout: 4x2
- Maximum speed: 38km/h (24mph)
- Range (on road): 300km (186 miles)

Autocarro Sahariano SPA AS 37

An AS 37 from the first series bearing Superflex Sigillo Verde tyres with the classic diamond pattern tread and disc wheels with lightening holes. (Fiat, Viberti)

The interior of the cargo bed. (Fiat, Viberti)

Developmental and Service History

The SPA AS 37 light truck, conceived in 1937, was derived from the TL 37 light tractor and was designed specifically for operations in North Africa. The impetus for its development is attributed to Maresciallo di Campo Italo Balbo, who was Governor-General of Libya, around 1937. The letters AS stood for *Autocarro Sahariano* (Saharan truck), and the vehicle was commonly referred to in Italian simply as the *Sahariano*. As well as the difference in body style compared to the TL 37 tractor, the AS 37 had improved filters, additional fuel and drinkable water tanks and more suitable Superflex Sigillo Verde tyres. A definitive version of the truck appeared in 1939.

Production was somewhat modest, with 1,291 examples totally or partially manufactured before 1942. Production data mixed TL 37s and AS 37s, but 802 trucks were in service on 30 April 1943, with a further 87 awaiting delivery. Following the 8 September 1943 Armistice, the Germans ordered a further 604 vehicles.

A number of SPA AS 37s were employed by the *Compagnie Auto-Sahariana* of the *Battaglione Sahariano*, a specialist unit that combined land and air personnel and vehicles. The *Battaglione* changed its composition between its establishment in 1938 and 1940, on the eve of Italy's entry into the Second World War. In addition to a HQ and a HQ detachment (equipped with light vehicles and machine guns), there was an *Avio-Sahariana Sezione* (Saharan Air Section) with Caproni Ca.309 Ghibli a reconnaissance and ground-attack aircraft; four *Auto-Sahariane Compagnie* (also with light vehicles and machine guns, with liaison, reconnaissance and defence tasks for oases and runways); and a *Meharista Compagnie* (camel cavalry). At the end of

An example equipped with canvas top and Superflex Artiglio tyres, except the spare wheel which is still a Superflex Sigillo Verde. (Fiat, Viberti)

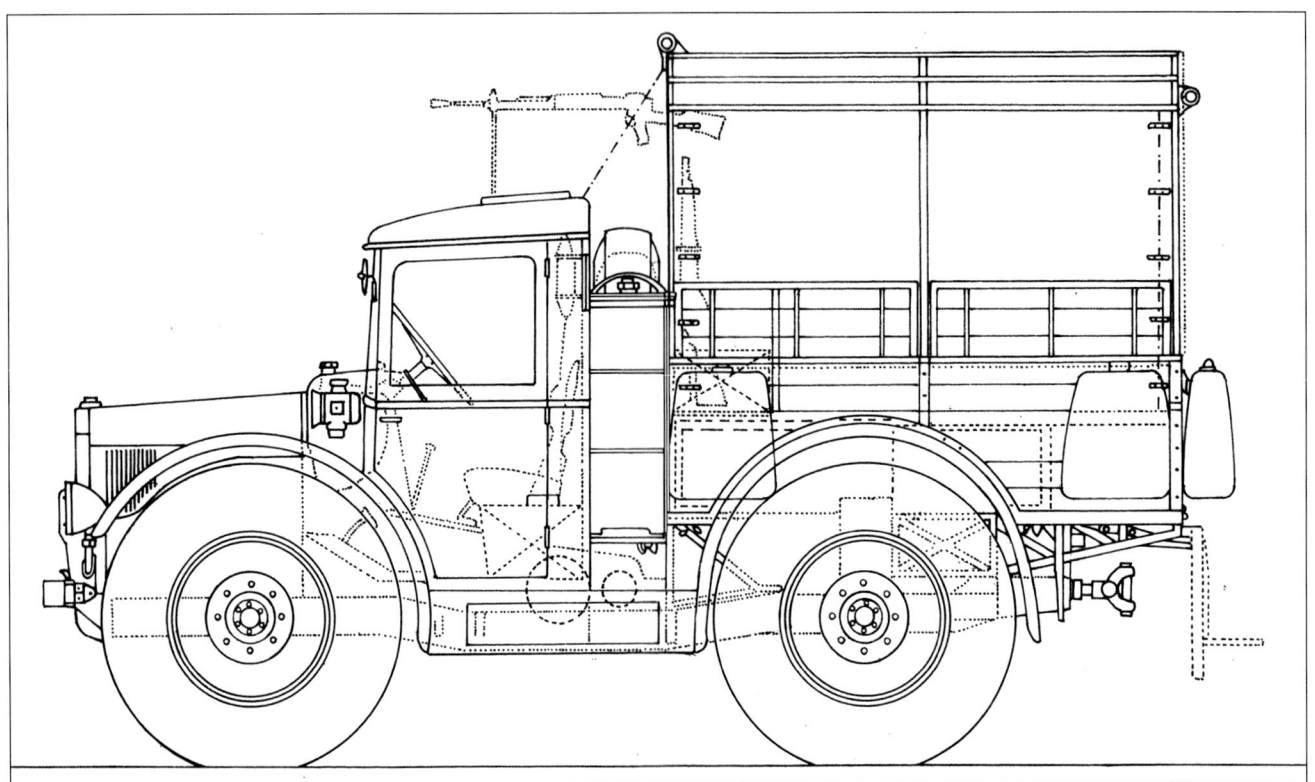

The Autocarro Sahariano AS 37 first series. (Drawing by R. Ciuffoletti and A. M. Feller – GMT)

1942, the military command of the Libyan Sahara converted two AS 37 trucks by eliminating the upper part of their cabs and mounting a 20mm Breda cannon on one example, and a 47/32 gun on the other. During the war, the companies also faced raids by the British Long Range Desert Group.

The AS 37 remained in service with the post-war Italian Army as an artillery tractor for only a brief period, given the wide availability of vehicles of Allied origin.

Technical Description

The AS 37 was laid out conventionally being a 4x4 vehicle with right-hand drive. The cab was all metal and had full doors with windows. The final version had five fuel tanks (one main tank containing 100 litres, two additional tanks with 100 litres each in the cargo bed and another two 50 litre tanks on the roof of the cab) plus four water tanks each containing 50 litres.

The wooden cargo body could carry a full rifle squad of eight men, or a 1,200kg payload and could be covered with a canvas cover; the bed had two seating benches, one each side, which had seats that could be raised to allow access to storage lockers

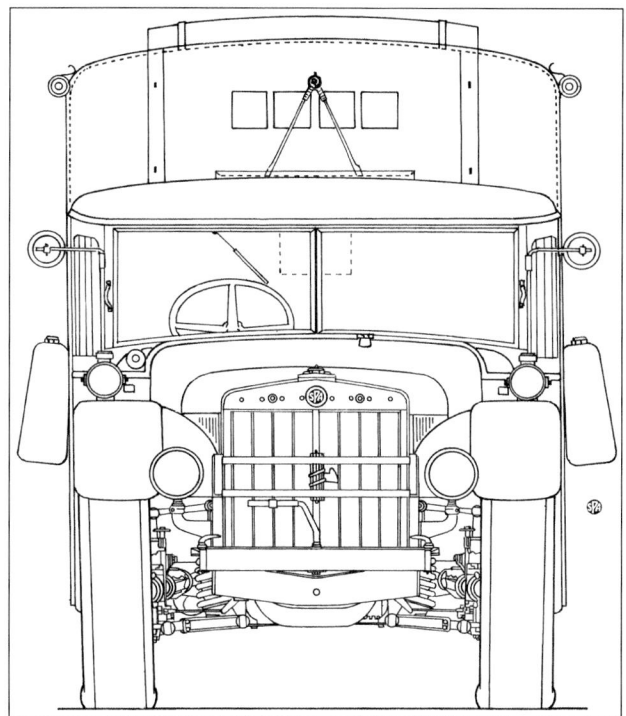
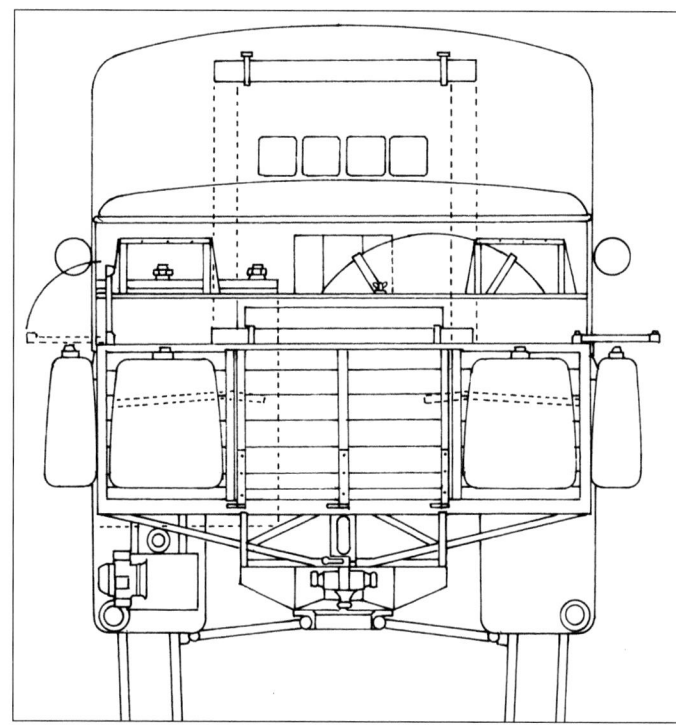

The Autocarro Sahariano AS 37 first series. (Drawing by R. Ciuffoletti and A. M. Feller – GMT)

One of the first examples of AS 37 light truck, at the factory and during trials in North Africa with provisional Superflex Sigillo Raiflex tyres. Note the doors with air vents. (Fiat, Viberti)

An AS 37 from the second series, with standard Sigillo Verde tyres, being tested in the Viberti factory in Turin. (Viberti)

142 ITALIAN SOFT-SKINNED VEHICLES OF THE SECOND WORLD WAR

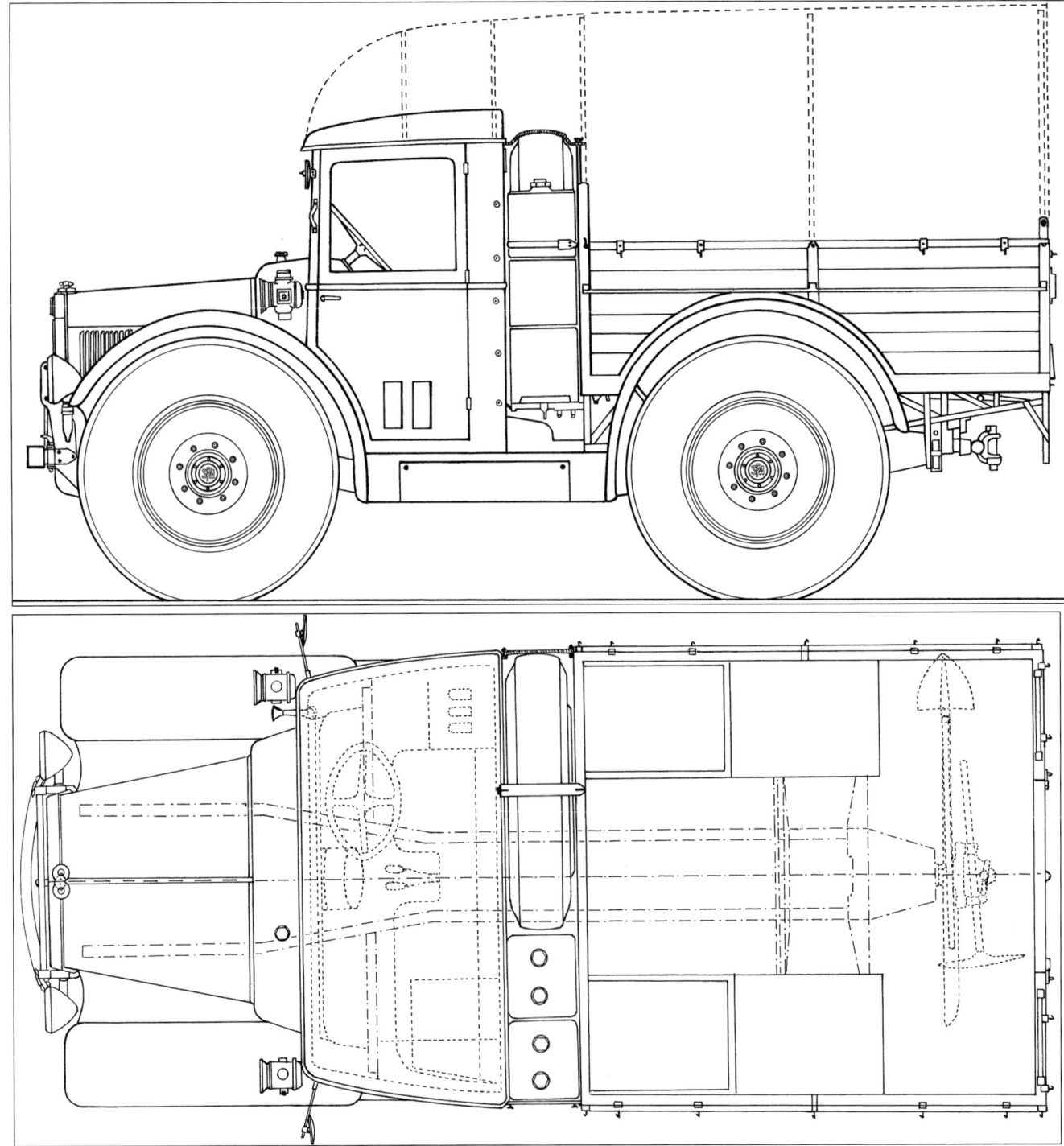

The Autocarro Sahariano AS 37 second series. (Drawing by A. M. Feller – GMT)

beneath. The tailgate consisted of the middle third of the rear panel and, when lowered, had a small integral step to allow easier access to the bed. Early versions of the AS 37 had the same wheels and tyres as the tractor (stamped steel wheels with 8 holes mounting Superflex Artiglio pneumatic tyres), while later production switched to pressed steel wheels with no holes, fitted with Superflex Sigillo Verde or Sigillo Verde Raiflex balloon tyres. A few had pressed steel wheels with Celerflex semi-pneumatic tyres.

The low pressure Superflex tyres enabled it to move easily over desert sand without bogging down. A tyre inflating compressor was carried and a spare tyre was mounted behind the driver's side of the cab. The truck had special oil filters for the engine and transmission. The suspension system was of the type with independent wheels; the front suspension included coil springs with hydraulic shock absorbers, while the rear one consisted of an inverted transversal semi-elliptical leaf spring, more robust compared to the TL 37.

The engine was the four-cylinder Fiat 18T, which was also used on the TL 37 artillery tractor. The manual transmission had four forward and one reverse gears.

A column of AS 37s belonging to a *Compagnia Auto-Sahariana* (Saharan Motorised Company), of the *Battaglione Sahariano* (Saharan Battalion), whose winged lion emblem is visible on the doors. (ACS)

Column of a *Compagnia Auto-Sahariana*.

One of the two AS 37 trucks modified by the *Comando del Sahara Libico* (Libyan Sahara Command or Libyan Sahara HQ) workshop; this example mounted a 47/32 gun. In the background, an AS42 Sahariana.

The second AS 37 modified by the *Comando del Sahara Libico* (Libyan Sahara Command or Libyan Sahara HQ) workshop mounting a Breda 20/65 cannon.

An artillery shell explodes near an AS 37 used as an observation platform.

An AS 37 radio van used by a *Compagnia Marconisti Motorizzati* (Motorised Wireless Operators Company). The van and trailer carried a radio station complete with equipment, antenna and personnel. (ACS)

A radio van with the antenna system deployed. (ACS)

Variants

Variants included a recovery vehicle, a bowser, a mobile repair shop, a closed body radio van and a field version mounting a ladder that could be raised on the rear of the bed and which was used as an artillery observation platform.

A derivative of the AS 37 was the Camionetta Desertica 43 derived from the field experiences in Libya with two modified AS 37 trucks; their closed cab, side doors and windshield were removed and the cargo body of one example was fitted with the 20mm Breda mod. 35 cannon, the second vehicle with the 47/32 gun.

Specifications

- Designation: Autocarro Sahariano SPA AS 37
- Producer: Società Piemontese Automobili (SPA), Turin
- Years produced: 1939–1944
- Number produced: Approximately 1,300
- Length: 4,670mm (15ft 4in)

- Width: 2,020mm (6ft 7in)
- Height: 2,650mm (8ft 8in)
- Unladen weight: 3,770kg (8,311lbs)
- Carrying capacity: 1,200kg (2,646lbs)
- Wheelbase: 2,500mm (8ft 2in)
- Front track: 1,518mm (5ft) with Superflex Artiglio tyres; 1,574mm (5ft 2in) with Superflex Sigillo Verde tyres
- Rear track: 1,518mm (5ft) with Superflex Artiglio tyres; 1,574mm (5ft 2in) with Superflex Sigillo Verde tyres
- Minimum turning radius: 5,000mm (16ft 5in)
- Minimum clearance: 390mm (1ft 3in)
- Fording depth: 700mm (2ft 3in)
- Bed, length: 2,000mm (6ft 7in)
- Bed, width: 1,900mm (6ft 3in)
- Bed, height: 600mm (1ft 11½in)
- Tyres: Celerflex 160 x 88; Superflex Artiglio 9.75 x 24; Superflex Sigillo Verde 11.25 x 24
- Engine: Fiat 18 TL, four-cylinder, water-cooled, 4,053cc, 52 HP @2,000 rpm
- Fuel: Gasoline
- Transmission: four speeds forward, one reverse
- Fuel capacity: 100 litres (26.4 US gallons, 22 Imperial gallons) plus 300 litres (79.25 US gallons, 66 Imperial gallons) in four supplemental tanks
- Drive layout: 4x4
- Speed: 50km/h (31mph)
- Range (on road): 870km (541 miles) with supplemental fuel tanks or canisters

This unidentified vehicle could be the prototype of a repair shop, based on the AS 37 second series but never entered production.